Discriminative Learning for Speech Recognition

Theory and Practice

Discriminative Learning for Speech Recognition: Theory and Practice
Xiadong He and Li Deng

ISBN: 978-3-031-01429-1 paperback

ISBN: 978-3-031-02557-0   ebook

DOI: 10.1007/978-3-031-02557-0

A Publication in the Springer series

SYNTHESIS LECTURES ON SPEECH AND AUDIO PROCESSING #4

Lecture #4

Series Editor: B. H. Juang, Georgia Institute of Technology

Series ISSN
ISSN 1932-121X   print
ISSN 1932-1678   electronic

Discriminative Learning for Speech Recognition

Theory and Practice

Xiaodong He and Li Deng
Microsoft Research

SYNTHESIS LECTURES ON SPEECH AND AUDIO PROCESSING #4

ABSTRACT

In this book, we introduce the background and mainstream methods of probabilistic modeling and discriminative parameter optimization for speech recognition. The specific models treated in depth include the widely used exponential-family distributions and the hidden Markov model. A detailed study is presented on unifying the common objective functions for discriminative learning in speech recognition, namely maximum mutual information (MMI), minimum classification error, and minimum phone/word error. The unification is presented, with rigorous mathematical analysis, in a common rational-function form. This common form enables the use of the growth transformation (or extended Baum–Welch) optimization framework in discriminative learning of model parameters. In addition to all the necessary introduction of the background and tutorial material on the subject, we also included technical details on the derivation of the parameter optimization formulas for exponential-family distributions, discrete hidden Markov models (HMMs), and continuous-density HMMs in discriminative learning. Selected experimental results obtained by the authors in firsthand are presented to show that discriminative learning can lead to superior speech recognition performance over conventional parameter learning. Details on major algorithmic implementation issues with practical significance are provided to enable the practitioners to directly reduce the theory in the earlier part of the book into engineering practice.

KEYWORDS

Speech recognition, discriminative learning, optimization, growth transformation, hidden Markov model, exponential-family distribution

Contents

1. **Introduction and Background** 1
   1.1 What is Discriminative Learning? 1
   1.2 What is Speech Recognition? 2
   1.3 Roles of Discriminative Learning in Speech Recognition 4
   1.4 Background: Basic Probability Distributions 5
       1.4.1 Multinomial Distribution 6
       1.4.2 Gaussian and Mixture-of-Gaussian Distributions 7
       1.4.3 Exponential-Family Distribution 7
   1.5 Background: Basic Optimization Concepts and Techniques 17
       1.5.1 Basic Definitions 18
       1.5.2 Necessary and Sufficient Conditions for an Optimum 18
       1.5.3 Lagrange Multiplier Method for Constrained Optimization 19
       1.5.4 Gradient Descent Method 20
       1.5.5 Growth Transformation Method: Introduction 21
   1.6 Organization of the Book 23

2. **Statistical Speech Recognition: A Tutorial** 25
   2.1 Introduction 25
   2.2 Language Modeling 26
   2.3 Acoustic Modeling and HMMs 27

3. **Discriminative Learning: A Unified Objective Function** 31
   3.1 Introduction 31
   3.2 A Unified Discriminative Training Criterion 32
       3.2.1 Notations 32
       3.2.2 The Central Result 32
   3.3 MMI and its Unified Form 33
       3.3.1 Introduction to MMI Criterion 33
       3.3.2 Reformulation of the MMI Criterion into Its Unified Form 34

3.4   MCE and its Unified Form 35
    3.4.1   Introduction to the MCE Criterion 35
    3.4.2   Reformulation of the MCE Criterion Into its Unified Form 38
3.5   Minimum Phone/Word Error and its Unified Form 39
    3.5.1   Introduction to the MPE/MWE Criterion 39
    3.5.2   Reformulation of the MPE/MWE Criterion Into
          Its Unified Form 40
3.6   Discussions and Comparisons 41
    3.6.1   Discussion and Elaboration on the Unified Form 41
    3.6.2   Comparisons With Another Unifying Framework 43

4.   Discriminative Learning Algorithm for Exponential-Family Distributions 47
4.1   Exponential-Family Models for Classification 47
4.2   Construction of Auxiliary Functions 48
4.3   GT Learning for Exponential-Family Distributions 49
4.4   Estimation Formulas for Two Exponential-Family Distributions 54
    4.4.1   Multinomial Distribution 54
    4.4.2   Multivariate Gaussian Distribution 55

5.   Discriminative Learning Algorithm for Hidden Markov Model 59
5.1   Estimation Formulas for Discrete HMM 59
    5.1.1   Constructing Auxiliary Function $F(\Lambda; \Lambda')$ 60
    5.1.2   Constructing Auxiliary Function $V(\Lambda; \Lambda')$ 60
    5.1.3   Simplifying Auxiliary Function $V(\Lambda; \Lambda')$ 61
    5.1.4   GT by Optimizing Auxiliary Function $U(\Lambda; \Lambda')$ 65
5.2   Estimation Formulas for CDHMM 67
5.3   Relationship with Gradient-Based Methods 70
5.4   Setting Constant D for GT-Based Optimization 71
    5.4.1   Existence Proof of Finite D in GT Updates for CDHMM 72

6.   Practical Implementation of Discriminative Learning 75
6.1   Computing $\Delta\gamma(i, r, t)$ in Growth-Transform Formulas 75
    6.1.1   Product Form of $C(s)$ (for MMI) 76
    6.1.2   Summation Form of $C(s)$ (MCE and MPE/MWE) 78
6.2   Computing $\Delta\gamma(i, r, t)$ Using Lattices 79
    6.2.1   Computing $\Delta\gamma(i, r, t)$ for MMI Involving Lattices 80
    6.2.2   Computing $\Delta\gamma(i, r, t)$ for MPE/MWE Involving Lattices 83
    6.2.3   Computing $\Delta\gamma(i, r, t)$ for MCE Involving Lattices 87

6.3   Arbitrary Exponent Scaling in MCE Implementation 88
6.4   Arbitrary Slope in Defining MCE Cost Function 89

7.   **Selected Experimental Results** 91
7.1   Experimental Results on Small ASR Tasks TIDIGITS 91
7.2   Telephony LV-ASR Applications 92

8.   **Epilogue** 97
8.1   Summary of Book Contents 97
8.2   Summary of Contributions 98
8.3   Remaining Theoretical Issue and Future Direction 99

Major Symbols Used in the Book and Their Descriptions 103

Mathematical Notation 105

Bibliography 107

Author Biography 111

CHAPTER 1

Introduction and Background

1.1 WHAT IS DISCRIMINATIVE LEARNING?

Discriminative learning is one of two major paradigms in constructing probabilistic pattern classifiers and recognizers, where classifiers usually deal with nonsequential data (i.e., with fixed-dimension input features) and the classification target is one of a limited set of categories, whereas recognizers handle sequential data (i.e., with variable-dimension input features) and the recognition target is an open output that can be of variable length. The first major paradigm is generative modeling and learning, which establishes and learns a model of the joint probability of the features and the class identity. In contrast, discriminative methods that characterize the second major paradigm either directly model the class posterior probability, or learn the parameters of the joint-probability model discriminatively so as to minimize classification/recognition errors.

The main purpose of this book is to present an extensive account on the basic ideas behind the approaches and techniques on discriminative learning, especially those that discriminatively learn the parameters of joint-probability models [e.g., hidden Markov models (HMMs)]. In addition, we also desire to position our treatment of the related algorithms in a wider context of learning and building statistical classifiers/recognizers from a more general context of machine learning. The Bayes decision theory serves as the basic formalism of the classification and recognition processes for achieving the optimal decision boundaries. Hence, the goal of pattern recognition can be described as finding the parameters of the classifiers or recognizers that minimize the error rate by using the available training samples. Within the two main paradigms for designing and learning statistical classifiers/recognizers, the generative paradigm uses the joint-probability model to perform the decision-making task based on the posterior probability of the class computed by Bayes rule [11, 43, 57]. The standard approach to learning (i.e., estimating) a generative model is maximum likelihood (ML). ML learning is considered a nondiscriminative approach because it aims at modeling the data distribution instead of directly separating class categories.

On the other hand, the discriminative classifiers/recognizers typically bypass the stage of building the joint-probability model while directly using the class posterior probability. This is exemplified by the celebrated argument that "one should solve the (classification/recognition) problem directly and never solve a more general problem as an intermediate step" [48]. This recognizer design philosophy is the basis of a wide range of popular machine learning methods including support vector machine [48], conditional random field [28, 37], and maximum entropy Markov models [15, 30], etc., where the "intermediate step" of estimating the joint distribution has been avoided. For example, in the recently proposed structured classification approach [15, 28, 30, 37] in machine learning and speech recognition, some well-known deficiencies of the HMM are addressed by applying "direct" discriminative learning, replacing the need for a probabilistic generative model by a set of flexibly selected, overlapping "features." Because the conditioning is made on the feature sequence and the "features" can be designed with long-contextual-span properties, the conditional-independence assumption made in the HMM is conceptually alleviated — provided that proper "features" can be constructed. How to design such features is a challenging research direction and it becomes a critical factor for the potential success of the structured discriminative approach, which departs from the "generative" component or joint distribution. On the other hand, local features can be much more easily designed that are appropriate for the generative approach, and many effective local features have been well established (e.g., cepstra, filter-bank outputs, etc. [10, 43]). Despite the high complexity of estimating joint distributions when the sole purpose is discrimination, the generative approach also has important advantages of facilitating knowledge incorporation and of conceptually straightforward analyses of classifier/recognizer components and their interactions.

Analyses of the capabilities and limitations associated with the two general machine learning paradigms discussed above lead to a practical pattern recognition framework that will be pursued in this book. That is, we attempt to establish a simplistic joint-distribution or generative model, with the complexity lower than what is required to accurately "generate" samples from the true distribution. To make such low-complexity generative models discriminate well, it requires parameter learning methods that are discriminative in nature so as to overcome the limitations in the simplistic model structures. This is in contrast to the generative approach of fitting the intraclass data as the conventional ML-based methods intend to accomplish. This type of practical framework has been applied to and guiding much of the recent work in speech recognition research, where HMMs are used as the low-complexity joint distribution for the local acoustic feature sequences of speech and the corresponding underlying linguistic label sequences (sentences, words, or phones, etc.).

1.2   WHAT IS SPEECH RECOGNITION?

Speech recognition is the process and the related technology for converting a speech signal into a sequence of words (or other linguistic units) by means of an algorithm implemented as a computer program. Speech recognition applications that have emerged over the last few years include voice dialing, call routing, interactive voice response, voice search, data entry and dictation, command and control (voice user interface with the computer), hands-free computing (automotive applications), structured document creation (e.g., medical and legal transcriptions), appliance control by voice, computer-aided language learning, content-based spoken audio search, and robotics.

Modern general-purpose speech recognition systems are generally based on HMMs, which will be described in some detail in Chapter 2. One reason why HMMs are popular in speech recognition is that their parameters can be trained or learned automatically and the learning techniques are simple and computationally feasible to use. In speech recognition, to give a simple setup, we imagine that HMMs generate a sequence of multidimensional real-valued or symbolic/discrete acoustic features, each corresponding to about 10 msecs of speech waveform. The real-valued vectors (or the discrete symbols) often consist of cepstral coefficients (or their vector-quantized codes), which are obtained by taking a Fourier transform of a short-time window of speech and decorrelating the spectrum by using a cosine transform. The continuous-density (CD) HMMs usually have, in each state, a probability distribution of a mixture of diagonal-covariance Gaussians. Discrete HMMs usually have, in each state, a nonparametric discrete distribution. Each word or phone will have different output distributions that are trained or learned automatically. An HMM for a sequence of words or phonemes is constructed by concatenating the individual trained HMMs for the separate words and phones.

Major developments in the technology of speech recognition over the past 50 years have been elegantly summarized in a recent keynote presentation at International Conference on Acoustics, Speech, and Signal Processing; the slides of that presentation can be found in http://www.ewh.ieee.org/soc/sps/stc/News/NL0704/furui-icassp2007.pdf. This long period has witnessed the field of speech recognition proceeding from its infancy to its current coming of age. Although far from a "solved" problem, it now has a growing number of practical applications in many sectors. Further research and development will enable increasingly more powerful systems, deployable on a world-wide basis.

Let us summarize the major developments of speech recognition in four areas. First, in the infrastructure area, Moore's law, in conjunction with the constantly shrinking cost of memory, has been instrumental in enabling speech recognition researchers to develop and run increasingly complex systems. The availability of common speech corpora for speech system training, development, and evaluation, has been critical in creating systems of increasing capabilities. Speech is a highly variable signal, characterized by many factors, and thus large corpora are critical in modeling it well enough for automated systems to achieve proficiency. Over the years, these corpora have been created, annotated, and distributed to the worldwide community. The character of the recorded speech has progressed from limited, constrained speech materials to masses of progressively more realistic, spontaneous, and "found" speech. The development and adoption of rigorous benchmark evaluations and standards have also been critical in developing increasingly powerful and capable speech recognition systems.

Second, in the area of knowledge representation, major advances in speech signal representations have included perceptually motivated acoustic features of speech. Architecturally, the most important development has been the searchable unified graph representations allowing multiple sources of knowledge to be incorporated in a common probabilistic framework.

Third, in the area of modeling and algorithms, the most significant paradigm shift has been the introduction of statistical methods, especially of the HMM method. More than 30 years after the initial use of HMMs in 1970s, this methodology still predominates. The ML-based expectation–maximization (EM) algorithm and the forward–backward or Baum–Welch algorithm have remained one of the principal means by which the HMMs are trained from data. In the area of language modeling, *N*-gram models have proved remarkably powerful and resilient despite their simplicity. Decision trees have been widely used to categorize sets of features, such as pronunciations from training data. Statistical discriminative learning techniques form the recent major innovations in speech recognition algorithms, which will be elaborated below and be the focus of the remainder of this book.

Fourth, in the area of recognition hypothesis search, key decoding or search strategies, originally developed in nonspeech applications, have focused on stack decoding (A* search), Viterbi, *N*-best, and lattice search/decoding. Derived originally from communications and information theory, stack decoding was subsequently applied to speech recognition systems. Viterbi or dynamic-programming based search is at present broadly applied to search alternative recognition hypotheses in virtually all modern speech recognition systems.

1.3   ROLES OF DISCRIMINATIVE LEARNING IN SPEECH RECOGNITION

As we just highlighted above, statistical discriminative learning has become a major theme in recent speech recognition research (e.g., [8, 9, 12, 18, 25, 31, 36, 37, 40, 42]). In particular, much of the striking progress in large-scale automatic speech recognition over the past few years has been attributed to the successful development and applications of discriminative learning (e.g., [31, 33, 40, 41]). Although the ML-based learning algorithm (i.e., the Baum–Welch algorithm) has been highly efficient and practical, it limits the performance of speech recognition. This is because ML learning relies on the assumption that the correct functional form of the joint probability between the data and the class categories is known and that there are sufficient and representative training data, both of which are often not realistic in practice. In the case of speech recognition, the data are speech feature sequences and the class categories are word sequences. As we discussed earlier, the currently most popular functional form of the probability model for speech is the HMM. Given the knowledge gained from many years of research in speech science, the assumptions made by the HMM are in many ways incorrect for the realistic processes in human speech. This inconsistency motivates the development of discriminative learning methods for speech recognition and high-lights their critical roles in improving speech recognition performance beyond the conventional ML-based learning techniques. The essence of discriminative learning as presented in this book is to learn the parameters of distribution models (e.g., HMMs) in such a way that the recognition errors or some measures of them are minimized directly via efficient and effective optimization techniques.

Two central issues in the development of discriminative learning methods for sequential pattern recognition and in particular for speech recognition are: (1) construction of the objective functions for optimization and (2) actual optimization techniques. There have been a wide variety of methods reported in the literature related to both of these issues (e.g., [8, 14, 18, 25, 31, 33, 34, 38, 42, 44, 46, 49]); however, their relationships have not been adequately understood. Because of the practical and theoretical importance of this problem, there is a pressing need for a unified ac-count of the numerous discriminative learning techniques in the literature. This book aims to fulfill this need while providing insights into the discriminative learning framework for sequential pattern classification and for speech recognition. In presenting discriminative learning in this chapter, we intend to address the issues of how the various discriminative learning techniques are related to and distinguished from each other, and what may be a deeper underlying scheme that can unify various ostensibly different techniques. Although the unifying review provided in this book is on a general class of pattern recognition problems associated with sequential characteristics, we will focus most of the discussions on those related to speech recognition and to the HMM [10, 43, 47]. We note that the HMM as well as the various forms of discriminative learning have been used in many signal processing areas beyond speech; for example, in bioinformatics [5, 13], in text and image classification/recognition [29, 53, 56], in video object classification [54], in natural language processing [7, 9], and in telerobotics [55]. It is our hope that the unifying review and the insights provided in this book will foster more principled and successful applications of discriminative learning in a wide range of signal processing disciplines, speech processing or otherwise.

1.4 BACKGROUND: BASIC PROBABILITY DISTRIBUTIONS

In this section, we provide the mathematical background for several basic probability distributions that will be used directly or as building blocks for more complex distributions in the remaining chapters of this book. The basic probability distributions discussed first will include multinomial distribution (discrete), Gaussian, and mixture-of-Gaussian distributions (continuous). Then we will present a more general form of the distributions, exponential-family distributions, which sub-sume a large number of discrete and continuous distributions. The more complex distributions (e.g., HMMs) built from the basic distributions will be presented in subsequent chapters.

1.4.1 Multinomial Distribution

Frequently, we need to handle discrete random variables that may take one of K possible values. Among all possible ways to express such variables, there is a convenient representation that the variable is represented by a K-dimensional vector x in which one of the elements $x(k)$ is equal to 1, and all other elements are equal to 0. For example, if we have a variable that can take $K = 6$ possible values and a particular observation of the variable happens to correspond to the third value, that is, $x(3) = 1$, then x can be represented by

$$x = [\, 0, 0, 1, 0, 0, 0 \,]^{\mathrm{T}}$$

If we denote the probability of $x(k) = 1$ by the parameter v_k, then the distribution of x is given by

$$p(x|v) = \prod_{k=1}^{K} v_k^{x(k)} \tag{1.1}$$

where $v = [v_1, \ldots, v_K]^{\mathrm{T}}$ is the parameter vector. Because v_k is a probability, that is, it is the probability of that the random variable takes the kth value, $\{v_k\}$ are constrained to satisfy $v_k \geq 0$ and $\sum_k v_k = 1$.

Now consider a data set X of N independent observations $x_1, \ldots, x_N$. If we denote the counts of observations at state k by the value m_k, we can have the joint distribution of the quantities $m_1, \ldots, m_K$, conditioned on the parameter vector v, and the total number of observations N, which takes the following form:

$$p(m_1, \ldots, m_K | v, N) = \binom{N}{m_1, \ldots, m_K} \prod_{k=1}^{K} v_k^{m_k} \tag{1.2}$$

This is known as the multinomial distribution. The normalization coefficient is the number of ways of partitioning N objects into K groups of size $m_1, \ldots, m_K$ and is computed as

$$\binom{N}{m_1, \ldots, m_K} = \frac{N!}{m_1! m_2! \ldots m_K!}$$

where the variables m_k are subject to the constraint

$$m_k \geq 0 \quad \text{and} \quad \sum_k m_k = N.$$

Note that (1.1) is a special case of the multinomial distribution for a single observation; that is, $N = 1$.

1.4.2 Gaussian and Mixture-of-Gaussian Distributions

The Gaussian or normal distribution is a widely used model for the distribution of continuous variables. When the random variable x is a scalar, the Gaussian probability density function (PDF) is

$$p(x|\lambda) = \frac{1}{(2\pi\sigma^2)^{1/2}} \exp\left\{ -\frac{1}{2}\frac{(x-\mu)^2}{\sigma^2} \right\} = N(x;\mu,\sigma^2) \qquad (1.3)$$

where the parameter set λ includes μ (mean) and σ (standard deviation). For a D-dimensional vector x, the multivariate Gaussian PDF takes the form of

$$p(x|\lambda) = \frac{1}{(2\pi)^{\frac{D}{2}}|\Sigma|^{\frac{1}{2}}} \exp\left\{ -\frac{1}{2}(x-\mu)^{\mathrm{T}}\Sigma^{-1}(x-\mu) \right\} = N(x;\mu,\Sigma) \qquad (1.4)$$

where $\lambda = \{\mu, \Sigma\}$ includes μ (mean vector) and Σ (covariance matrix). The Gaussian distribution is commonly used in many engineering and science disciplines including speech processing. The popularity arises not only from its highly desirable computational properties, but also from its ability to approximate many naturally occurring real-world data due to the central limit theorem.

Mixture-of-Gaussian distributions. Unfortunately, in some speech processing problems including speech recognition, the Gaussian distribution is inadequate. The inadequacy comes from its unimodal property, whereas most speech features have multimodal distributions. An appropriate distribution is the following mixture-of-Gaussian distribution with the desirable multimodal property:

$$p(x|\lambda) = \sum_{m=1}^{M} c_m N(x;\mu_m,\sigma_m^2) \qquad (1.5)$$

where the variable x is a scale and $\lambda = \{c_m, \mu_m, \sigma_m^2; m = 1, 2 ..., M\}$ or

$$p(x|\lambda) = \sum_{m=1}^{M} c_m N(x;\mu_m,\Sigma_m)$$

where the variable x is a vector and $\lambda = \{c_m, \mu_m, \Sigma_m; m = 1, 2 ..., M\}$.

1.4.3 Exponential-Family Distribution

Both multinomial and Gaussian distributions (but not the mixture-of-Gaussians) discussed above are special cases of a broad class of distributions known as the exponential family, including both continuous and discrete distributions. This general family of distributions is defined by the following PDF:

$$p(x|\theta) = h(x) \cdot \exp\left\{\theta^{\mathrm{T}} T(x) - A(\theta)\right\} \tag{1.6}$$

where x can be scalar or vector, and may be discrete or continuous. Here, θ is called the natural parameters of the distribution, $T(x)$ is some function of x, $A(\theta)$ is the cumulative generating function, and $h(x)$ is the base measure, which is a function of x. To obtain a normalized distribution, we need to take integration of both sides of (1.6) and set it to one:

$$\int p(x|\theta)\mathrm{d}x = \exp\left(-A(\theta)\right) \int h(x) \cdot \exp\left(\theta^{\mathrm{T}} T(x)\right) \mathrm{d}x = 1 \tag{1.7}$$

Therefore,

$$\exp\left(A(\theta)\right) = \int h(x) \cdot \exp\left(\theta^{\mathrm{T}} T(x)\right) \mathrm{d}x \tag{1.8}$$

For a discrete random variable x, the integration above should be replaced by summation.

Convexity of the exponential-family distribution. Let us first consider the properties of $A(\theta)$. Examine the first-order derivative of $A(\theta)$. Taking the gradient of both side of (1.8) with respect to θ, we have

$$\exp[A(\theta)] \cdot \nabla A(\theta) = \int h(x) \cdot \exp\left[\theta^{\mathrm{T}} T(x)\right] \cdot T(x)\mathrm{d}x$$

Rearranging and making use of (1.6), we obtain

$$\nabla A(\theta) = \exp[-A(\theta)] \cdot \int h(x) \cdot \exp\left[\theta^{\mathrm{T}} T(x)\right] \cdot T(x)\mathrm{d}x = \mathbb{E}_{p(x|\theta)}[T(x)] \tag{1.9}$$

After using the chain rule and the matrix derivative formula of $\nabla(f(\theta) \times a) = \nabla f(\theta) \times a^{\mathrm{T}}$, the second-order derivative of $A(\theta)$ can be obtained based on (1.9):

$$\nabla^2 A(\theta) = -\exp[-A(\theta)] \cdot \nabla A(\theta) \cdot \int h(x) \cdot \exp\left[\theta^{\mathrm{T}} T(x)\right] \cdot T(x)^{\mathrm{T}}\mathrm{d}x$$
$$+ \exp[-A(\theta)] \cdot \int h(x) \cdot \exp\left[\theta^{\mathrm{T}} T(x)\right] \cdot T(x) \cdot T(x)^{\mathrm{T}}\mathrm{d}x$$

Using (1.9) again, we have

$$\nabla^2 A(\theta) = -\mathbb{E}_{p(x|\theta)}[T(x)]\mathbb{E}_{p(x|\theta)}[T(x)]^{\mathrm{T}} + \mathbb{E}_{p(x|\theta)}\left[T(x)T(x)^{\mathrm{T}}\right] = \mathrm{Cov}_{p(x|\theta)}[T(x)] \succeq 0$$

That is, the second-order derivative of $A(\theta)$ is positive definite. Therefore, $A(\theta)$ is a convex function of θ.

Maximum likelihood estimation and sufficient statistic of the exponential-family distribution. Now, let us consider the problem of estimating the parameter vector $\boldsymbol{\theta}$ in the general exponential-family distribution (1.6) using the technique of maximum likelihood (ML). In maximum likelihood estimation, consider that there is a set of independent identically distributed data denoted by $X = \{x_1, \dots, x_n\}$, for which the likelihood function is given by

$$p(X|\theta) = \left(\prod_{n-1}^{N} h(x_n) \right) \cdot \exp\left[\theta^{\mathrm{T}} \sum_{n=1}^{N} T(x_n) - N \cdot A(\theta) \right]$$

Setting the gradient of $\ln(p(X|\theta))$ with respect to θ to zero, we obtain the following condition to be satisfied by the maximum likelihood estimate

$$\nabla A(\theta) = \frac{1}{N} \sum_{n=1}^{N} T(x_n) \qquad (1.10)$$

which can be solved to obtain θ_{ML}. Because $A(\theta)$ is convex, there is one global unique ML solution for θ_{ML}.

From (1.10), we observe that the solution to the ML estimate depends on the data only through $\sum_{n=1}^{N} T(x_n)$, which is therefore called sufficient statistic of the distribution (1.6). In computing the ML estimate, we only need to store the value of the sufficient statistic.

The above sufficiency property holds for discriminative learning, and we will defer the discussion to Chapter 4.

Exponential form of the multinomial distribution. It can be verified that the distributions discussed in the previous sections are members of the exponential family.

Let us consider the multinomial distribution that, for a single observation x, takes the form

$$p(x|v) = \prod_{k=1}^{K} v_k^{x(k)} \qquad (1.11)$$

There are parameter constraints of $v_k \geq 0$ and $\sum_{k=1}^{K} v_k = 1$ in this form of distribution, which is to be removed. Due to the sum-to-one constraint, there are a total of $K-1$ free parameters. For instance, v_K can be expressed by the remaining $K-1$ parameters through $v_k = 1 - \sum_{j=1}^{K-1} v_j$, thus leaving $K-1$ free parameters. Note that these remaining $K-1$ parameters are still subject to the constraints $v_k \geq 0$ and $\sum_{j=1}^{K-1} v_j \leq 1$, $k = 1, \dots, K-1$. Note also that $\sum_{k=1}^{K} x(k) = 1$. We now rewrite the distribution

$$\prod_{k=1}^{K} v_k^{x(k)} = \exp\left\{ \sum_{k=1}^{K} x(k) \ln v_k \right\}$$

$$= \exp\left\{ \sum_{k=1}^{K-1} x(k) \ln v_k + \left(1 - \sum_{k=1}^{K-1} x(k)\right) \ln \left(1 - \sum_{j=1}^{K-1} v_j\right) \right\}$$

$$= \exp\left\{ \underbrace{\sum_{k=1}^{K-1} x(k) \ln \left(\frac{v_k}{1 - \sum_{j=1}^{K-1} v_j}\right)}_{\theta^{\mathrm{T}} T(x)} + \underbrace{\ln\left(1 - \sum_{j=1}^{K-1} v_j\right)}_{-A(\theta)} \right\}$$

and then construct the $K - 1$ dimensional natural parameter vector $\theta = [\theta_1,\ldots,\theta_{k-1}]^{\mathrm{T}}$ such that

$$\theta_k = \ln\left(\frac{v_k}{1 - \sum_{j=1}^{K-1} v_j}\right) \tag{1.12}$$

After identifying the parameters (as well as the sufficient statistic) above in the standard form, we now need to express $A(\theta)$ in terms of the parameters of (1.12). To do this, we rewrite (1.12) as

$$\exp(\theta_k) = \frac{v_k}{1 - \sum_{j=1}^{K-1} v_j} \tag{1.13}$$

Summing both sides of (1.13) over k, we have

$$\sum_{k=1}^{K-1} \exp(\theta_k) = \frac{\sum_{k=1}^{K-1} v_k}{1 - \sum_{j=1}^{K-1} v_j}$$

After adding one on both sides, we obtain

$$1 + \sum_{k=1}^{K-1} \exp(\theta_k) = \frac{1}{1 - \sum_{j=1}^{K-1} v_j} \tag{1.14}$$

Then, substituting the right-hand side of (1.14) into (1.13), we have

$$v_k = \frac{\exp(\theta_k)}{1 + \sum_{j=1}^{K-1} \exp(\theta_j)} \quad \text{and} \tag{1.15}$$

$$\ln\left(1 - \sum_{j=1}^{K-1} v_j\right) = \ln\left(1 - \frac{\sum_{k=1}^{K-1} \exp(\theta_k)}{1 + \sum_{j=1}^{K-1} \exp(\theta_j)}\right) = -\ln\left(1 + \sum_{j=1}^{K-1} \exp(\theta_j)\right)$$

Therefore, comparing with the standard form (1.6) of the exponential family distribution, we identify:

$$h(x) = 1$$

$$T(x) = \tilde{x} = [x_1, \ldots, x_{K-1}]^{\mathrm{T}}$$

$$A(\theta) = \ln\left(1 + \sum_{j=1}^{K-1} \exp(\theta_j)\right)$$

where $\tilde{x}$ is an observation vector that only contains the first $K-1$ elements of x. Furthermore,

$$\frac{\partial A(\theta)}{\partial \theta} = \frac{1}{1 + \sum_{j=1}^{K-1} \exp(\theta_j)} \begin{bmatrix} \exp(\theta_1) \\ \vdots \\ \exp(\theta_{K-1}) \end{bmatrix} = \begin{bmatrix} v_1 \\ \vdots \\ v_{K-1} \end{bmatrix} = \tilde{v} \tag{1.16}$$

where we denote by $\tilde{v}$ the partial parameter vector that only contains the first $K-1$ parameters.

We now discuss ML parameter estimation. According to (1.10), the maximum likelihood estimation of θ should satisfy the following condition:

$$\tilde{v} = \frac{1}{N} \sum_{n=1}^{N} \tilde{x}_n \tag{1.17}$$

By summing both sides of (1.17) over $k = 1, \ldots, K-1$, we have

$$\sum_{k=1}^{K-1} v_k = \frac{1}{N} \sum_{n=1}^{N} \sum_{k=1}^{K-1} x_n(k)$$

Therefore, we have

$$v_K = 1 - \sum_{k=1}^{K-1} v_k = 1 - \frac{1}{N}\sum_{n=1}^{N}\sum_{k=1}^{K-1} x_n(k) = \frac{1}{N}\sum_{n=1}^{N}\left(1 - \sum_{k=1}^{K-1} x_n(k)\right) = \frac{1}{N}\sum_{n=1}^{N} x_n(K) \qquad (1.18)$$

Combining (1.17) and (1.18), we have the ML estimation formula for the multinomial distribution

$$v_{\mathrm{ML}} = \frac{1}{N}\sum_{n=1}^{N} x_n$$

Exponential form of the univariate Gaussian distribution. Let us consider the single-variable Gaussian distribution:

$$p(x|\lambda) = \frac{1}{(2\pi\sigma^2)^{1/2}}\exp\left\{-\frac{(x-\mu)^2}{2\sigma^2}\right\}$$

$$= \underbrace{\frac{1}{\sqrt{2\pi}}}_{h(x)}\exp\left\{\underbrace{-\frac{x^2}{2\sigma^2} + \frac{2x\mu}{2\sigma^2}}_{\theta^{\mathrm{T}}T(x)}\underbrace{-\frac{\mu^2}{2\sigma^2} - \ln(\sigma)}_{-A(\theta)}\right\}$$

Therefore, we identify:

$$h(x) = 1/\sqrt{2\pi}$$

$$T(x) = \left[x, x^2\right]^{\mathrm{T}}$$

$$\theta = [\theta_1, \theta_2]^{\mathrm{T}} = \left[\frac{\mu}{\sigma^2}, \frac{-1}{2\sigma^2}\right]^{\mathrm{T}} \qquad (1.19)$$

To express $A(\theta)$ in terms of the parameters in the form of (1.19), we rewrite (1.19) to obtain

$$\mu = \frac{\theta_1}{(-2\theta_2)}$$

$$\sigma = (-2\theta_2)^{-\frac{1}{2}}$$

And therefore

$$A(\theta) = \frac{\mu^2}{2\sigma^2} + \ln(\sigma) = -\frac{\theta_1^2}{4\theta_2} - \frac{1}{2}\ln(-2\theta_2)$$

Exponential form of the multivariate Gaussian distribution. Let us consider the D-dimensional multivariate Gaussian distribution:

$$p(x|\lambda) = \frac{1}{(2\pi)^{\frac{D}{2}}|\Sigma|^{\frac{1}{2}}} \exp\left\{ -\frac{1}{2}(x-\mu)^{\mathrm{T}}\Sigma^{-1}(x-\mu) \right\}$$

$$= \underbrace{\frac{1}{(2\pi)^{\frac{D}{2}}}}_{b(x)} \exp\left\{ \underbrace{-\frac{1}{2}x^{\mathrm{T}}\Sigma^{-1}x + \frac{1}{2}x^{\mathrm{T}}\Sigma^{-1}\mu + \frac{1}{2}\mu^{\mathrm{T}}\Sigma^{-1}x}_{\theta^{\mathrm{T}}T(x)} \underbrace{-\frac{1}{2}\mu^{\mathrm{T}}\Sigma^{-1}\mu - \frac{1}{2}\ln|\Sigma|}_{-A(\theta)} \right\}$$

Therefore, we can identify

$$b(x) = (2\pi)^{-\frac{D}{2}}$$

$$T(x) = \begin{bmatrix} x' \\ x'' \end{bmatrix}$$

$$\theta = \begin{bmatrix} \theta' \\ \theta'' \end{bmatrix}$$

where we denote by

$$x' = x$$
$$x'' = \mathrm{Vec}\left(xx^{\mathrm{T}}\right)$$

(1.20)

and

$$\theta' = \Sigma^{-1}\mu$$
$$\theta'' = \mathrm{Vec}\left(-\frac{1}{2}\Sigma^{-1}\right)$$

(1.21)

In the above, we define $\mathrm{Vec}(\bullet)$ as a function that converts a matrix into a column vector in the following manner: First, it concatenates rows of the matrix one by one to form a row vector, and then transport the result to a column vector. Correspondingly, we define the inverse function of $\mathrm{Vec}(\bullet)$ as $\mathrm{IVec}(\bullet)$, which converts a column vector to a matrix. For example,

$$\text{Vec}\left(\begin{bmatrix} a_{11} & a_{12} \\ a_{21} & a_{22} \end{bmatrix}\right) = \begin{bmatrix} a_{11} \\ a_{12} \\ a_{21} \\ a_{22} \end{bmatrix} \quad \text{and} \quad \text{IVec}\left(\begin{bmatrix} a_{11} \\ a_{12} \\ a_{21} \\ a_{22} \end{bmatrix}\right) = \begin{bmatrix} a_{11} & a_{12} \\ a_{21} & a_{22} \end{bmatrix}$$

We now express $A(\theta)$ as an explicit function of the parameter θ as defined in (1.21). For simplicity of notation, we define matrix Θ as

$$\Theta = \text{IVec}\left(\theta''\right)$$

Note Θ is a symmetric matrix; that is, $\Theta = \Theta^{\text{T}}$ and $\Theta^{-1} = (\Theta^{-1})^{\text{T}}$.

Then we obtain

$$\mu = -\frac{1}{2}\Theta^{-1}\theta'$$

$$\Sigma = -\frac{1}{2}\Theta^{-1} = (-2\Theta)^{-1}$$

And we can derive

$$A(\theta) = \frac{1}{2}\mu^{\text{T}}\Sigma^{-1}\mu + \frac{1}{2}\ln|\Sigma|$$

$$= \frac{1}{2}\frac{-1}{2}\theta'^{\text{T}}\Theta^{-1}\theta' - \frac{1}{2}\ln|-2\Theta|$$

$$= -\frac{1}{4}\theta'^{\text{T}}\Theta^{-1}\theta' - \frac{1}{2}\ln|-2\Theta|$$

We now discuss ML-based parameter estimation. Using matrix calculus, we obtain

$$\frac{\partial A(\theta)}{\partial \theta'} = -\frac{1}{2}\Theta^{-1}\theta' = \mu \tag{1.22}$$

$$\frac{\partial A(\theta)}{\partial \theta''} = \text{Vec}\left(\frac{\partial A(\theta)}{\partial \Theta}\right) = \text{Vec}\left(\frac{1}{4}\Theta^{-1}\theta'\theta'^{\text{T}}\Theta^{-1} - \frac{1}{2}(-2)(-2\Theta)^{-1}\right) = \text{Vec}\left(\mu\mu^{\text{T}} + \Sigma\right) \tag{1.23}$$

According to (1.10), the ML estimate of θ should satisfy the following condition:

$$
\begin{bmatrix} \dfrac{\partial A(\theta)}{\partial \theta'} \\[3mm] \dfrac{\partial A(\theta)}{\partial \theta''} \end{bmatrix} = \begin{bmatrix} \dfrac{1}{N}\displaystyle\sum_{n=1}^{N} x'_n \\[3mm] \dfrac{1}{N}\displaystyle\sum_{n=1}^{N} x''_n \end{bmatrix} \tag{1.24}
$$

Therefore, after substituting (1.20), (1.22), and (1.23) into (1.24), we obtain

$$
\begin{bmatrix} \mu \\[2mm] \mathrm{Vec}\left(\mu\mu^{\mathrm{T}} + \Sigma\right) \end{bmatrix} = \begin{bmatrix} \dfrac{1}{N}\displaystyle\sum_{n=1}^{N} x_n \\[3mm] \dfrac{1}{N}\mathrm{Vec}\left(\displaystyle\sum_{n=1}^{N} x_n x_n^{\mathrm{T}}\right) \end{bmatrix}
$$

After rearrangement and canceling out the Vec( ) function on both sides, we have the estimation formula:

$$
\mu_{\mathrm{ML}} = \frac{1}{N}\sum_{n=1}^{N} x_n
$$

$$
\Sigma_{\mathrm{ML}} = \frac{1}{N}\sum_{n=1}^{N} x_n x_n^{\mathrm{T}} - \mu_{\mathrm{ML}}\mu_{\mathrm{ML}}^{\mathrm{T}}
$$

In addition to the multinomial and Gaussian distributions that are commonly used in speech modeling, we here also introduce a few other members of the exponential family. As will be shown in the following chapters, discriminative training for the general exponential family distributions is applicable to all the distributions discussed here.

Exponential form of the Poisson distribution. Poisson distribution has the following conventional form for one dimensional discrete variable:

$$
p(x|\lambda) = \frac{1}{x!}\lambda^x \exp(-\lambda) \qquad\qquad x = 0,\ 1,\ 2,\ \ldots
$$

$$
= \underbrace{\frac{1}{x!}}_{b(x)}\exp\left\{\underbrace{x\ln(\lambda)}_{\theta^{\mathrm{T}}T(x)}\ \underbrace{-\lambda}_{-A(\theta)}\right\}
$$

Therefore, we identify the quantities in the standard form of the exponential family:

$$h(x) = 1/x!$$
$$T(x) = x$$
$$\theta = \ln(\lambda)$$
$$A(\theta) = \lambda = e^{\theta}$$

Exponential form of the exponential distribution. Exponential distribution has the conventional form:

$$p(x|\lambda) = \lambda \exp(-\lambda x) \qquad x \in \mathbb{R}^+$$

$$= \exp \left\{ \underbrace{-\lambda x}_{\theta^{\mathrm{T}} T(x)} + \underbrace{\ln(\lambda)}_{-A(\theta)} \right\}$$

from which we identify:

$$h(x) = 1$$
$$T(x) = x$$
$$\theta = -\lambda$$
$$A(\theta) = -\ln(\lambda) = -\ln(-\theta)$$

Exponential form of the Dirichlet distribution. Dirichlet distribution takes the following form

$$p(x|\alpha) = \frac{\Gamma\left(\sum_{k=1}^{K} \alpha_k\right)}{\prod_{k=1}^{K} \Gamma(\alpha_k)} \prod_{k=1}^{K} x(k)^{\alpha_k - 1}$$

where $\alpha = [\alpha_1,..., \alpha_K]^{\mathrm{T}}$ is the parameter vector, $\Gamma(\cdot)$ is the Gamma function defined as

$$\Gamma(z) = \int_0^{\infty} t^{z-1} e^{-z} dt \qquad (1.25)$$

and $x = [x(1), ..., x(K)]^{\mathrm{T}}$ is a K-dimensional observation vector with the constraints: $0 \leq x(k) \leq 1$, $\sum_{k=1}^{K} x(k) = 1$. Rewriting (1.25), we have

$$p(x|\alpha) = \exp\left\{ \ln\frac{\Gamma\left(\sum_{k=1}^{K}\alpha_k\right)}{\prod_{k=1}^{K}\Gamma(\alpha_k)} + \sum_{k=1}^{K}(\alpha_k - 1)\ln(x(k)) \right\}$$

$$= \underbrace{\exp\left\{ -\sum_{k=1}^{K}\ln(x(k)) \right\}}_{h(x)} \exp\left\{ \underbrace{\sum_{k=1}^{K}\alpha_k \ln(x(k))}_{\theta^{\mathrm{T}}T(x)} + \underbrace{\ln\frac{\Gamma\left(\sum_{k=1}^{K}\alpha_k\right)}{\prod_{k=1}^{K}\Gamma(\alpha_k)}}_{-A(\theta)} \right\}$$

from which we can identify:

$$h(x) = \exp\left\{ -\sum_{k=1}^{K}\ln(x(k)) \right\}$$

$$T(x) = \begin{bmatrix} \ln(x(k)) \\ \vdots \\ \ln(x(K)) \end{bmatrix}$$

$$\theta = \alpha$$

$$A(\theta) = -\ln\frac{\Gamma\left(\sum_{k=1}^{K}\theta_k\right)}{\prod_{k=1}^{K}\Gamma(\theta_k)}$$

1.5 BACKGROUND: BASIC OPTIMIZATION CONCEPTS AND TECHNIQUES

In this section, we provide the mathematical background for basic optimization concepts and pertinent techniques that will be used in the remaining chapters of this book. In particular, we will introduce the growth-transformation-based optimization technique that applies to specific, rational forms of object functions. All topics discussed in this section will be used as the basic material for the following chapters in this book.

1.5.1   Basic Definitions

Let a vector Λ be in a K-dimensional parameter space, $\Lambda \in R^K$, and let $O(\Lambda)$ be a real-valued function of Λ. When we want to optimize the function $O(\Lambda)$, we call it the objective function w.r.t. to the parameter set Λ.

The function $O(\Lambda)$ with its domain $\Lambda \in R^K$ is said to have a global minimum Λ^* if

$$O(\Lambda^*) \leq O(\Lambda)$$

for all $\Lambda \in R^K$. The function $O(\Lambda)$ is said to have a global maximum Λ^{**} if

$$O(\Lambda^{**}) \geq O(\Lambda)$$

for all $\Lambda \in R^K$.

The function $O(\Lambda)$ is said to have a local minimum Λ_0 if

$$O(\Lambda_0) \leq O(\Lambda)$$

for all Λ in the neighborhood of Λ_0.

Note that since

$$\min O(\Lambda) = - \left[\max \left(-O(\Lambda) \right) \right]$$

a minimization problem is equivalent to a maximization one. We thus will treat both of these problems as the same optimization problem.

The vector of partial derivatives of $O(\Lambda)$ w.r.t. Λ is called the gradient vector, which is often denoted by $\nabla O(\Lambda)$. The matrix of second-order partial derivatives of $O(\Lambda)$ is called the Hessian matrix, denoted by H_Λ.

1.5.2   Necessary and Sufficient Conditions for an Optimum

A necessary condition for a function $O(\Lambda)$ to have a local optimum at Λ^* is that the gradient vector has all zero components:

$$\nabla O(\Lambda^*) = 0$$

as long as $\nabla O(\Lambda)$ exists and is continuous at Λ^*. This necessary condition can be directly proved using Taylor series expansion.

Note that $\nabla O(\Lambda^*) = 0$ is only a necessary condition; that is, a point Λ^* satisfying $\nabla O(\Lambda^*) = 0$ may be just a stationary or saddle point, not an optimum point.

However, in many optimization problems including those in speech processing, previous knowledge about the nature of the objective function in the problem domain can eliminate the possibility of having a stationary point.

To theoretically guarantee an optimum point (i.e., elimination of the possibility of a stationary point), we have the following sufficient condition: Let there exist continuous partial derivatives up to the second order for objective function $O(\Lambda)$. If the gradient vector $\nabla O(\Lambda^*) = 0$ and the Hessian matrix H_A is positive definite, then Λ^* is a local minimum. Similarly, if the gradient vector $\nabla O(\Lambda^*) = 0$ and the Hessian matrix H_A is negative definite, then Λ^* is a local maximum.

Again, the proof of the above condition comes also directly from applying Taylor series expansion.

The necessary and sufficient conditions discussed above are applied to optimization problems with no constraints. For the situation where constraints must be imposed, the related optimization problems are discussed next.

1.5.3 Lagrange Multiplier Method for Constrained Optimization

The Lagrange multiplier method is a popular method in speech processing, as well as in many other optimization problems, which converts constrained optimization problems into unconstrained ones. It uses a linear combination of the objective function and the constraints to form a new objective function with no constraints.

The constrained optimization problem, where the constraints are in the form of equalities, can be formally described as follows: Find $\Lambda = \Lambda^*$ that optimizes the objective function $O(\Lambda)$ subject to the M constraints:

$$g_1(\Lambda) = b_1,$$
$$g_2(\Lambda) = b_2,$$
$$\dots$$
$$g_M(\Lambda) = b_M.$$

The Lagrange multiplier method solves the above problem by forming a new objective function for the equivalent unconstrained optimization:

$$F(\Lambda, \lambda) = O(\Lambda) + \sum_{m=1}^{M} \lambda_m \left[g_m(\Lambda) - b_m \right]$$

where $\lambda = (\lambda_1, \lambda_2, \dots, \lambda_M)$ are called Lagrange multipliers.

Optimization of the new objective function $F(\Lambda,\lambda)$ proceeds by setting its partial derivatives to zero with respect to each vector component of Λ and λ. This produces a set of $K + M$ equations that determine the $K + M$ unknowns including the desired solution $\Lambda = \Lambda^*$ for optimization.

When the constraints are in the form of inequalities, rather than of equalities as discussed above, a common method for optimization is to transform the related variables so as to eliminate the constraints. For example, if the constraint is $\Lambda > 0$ (e.g., as required for estimating the variance, which is always positive, in a PDF), then we can transform Λ into $\Lambda' = \exp(\Lambda)$. Because Λ and Λ' are monotonically related, optimization of one automatically gives the solution to the other. However, when using this type of transformation techniques, one should be aware of the sensitivity problem in the solution.

1.5.4   Gradient Descent Method

One popular family of numerical methods for optimization is based on gradient descent. As discussed earlier, the gradient is a vector in a K-dimensional space where the objective function is defined. The effectiveness of these gradient-based methods derives from its important property: The gradient vector represents the direction of steepest ascent of the objective function, and the negative gradient vector represents the direction of steepest descent. That is, if we move along the gradient direction from any point in the K-dimensional space over which the objective function is defined, then the function value increases at the fastest rate.

Note that the direction of steepest ascent is a local and not a global property. Hence, all the optimization methods based on gradients give only local optimum, and not global optimum. Due to the steepest ascent or descent property associated with the gradient vector, any method that makes use of it can be expected to find an optimum point faster than the methods without using it.

In the steepest descent method, one uses the negative gradient vector, $\nabla O(\Lambda)$, as a direction for minimizing an objective function $O(\Lambda)$. In this method, an initial point $\Lambda^{(0)}$ is supplied, and it iteratively moves toward the optimal point using the updating equation:

$$\Lambda^{(t+1)} = \Lambda^{(t)} - \alpha_{\min}^{(t)} \nabla O(\Lambda^{(t)})$$

where $\alpha_{\min}^{(t)}$ is called the step size, and in the strict steepest descent method, the step size is optimized along the search direction $\nabla O(\Lambda^{(t)})$. That is, in each iteration of steepest descent, $\alpha_{\min}^{(t)}$ is found that minimizes $O[\Lambda^{(t)} - \alpha_{\min}^{(t)} \nabla O(\Lambda^{(t)})]$. In practice, for large-scale optimization problems such as speech recognizer training, the above procedure is difficult and $\alpha_{\min}^{(t)}$ is often determined empirically.

1.5.5 Growth Transformation Method: Introduction

The gradient descent method discussed above can be applied to any objective function, as long as the gradient can be computed efficiently (analytically or numerically, especially analytically). There are many other optimization techniques that take advantage of higher-order gradients, such as Newton's method. However, if the objective function has a special structure, more efficient optimization techniques than the gradient-based ones can be used. In this section, we provide preliminaries to optimizing rational functions, a common type of structure in the objective function, by a nongradient-based technique called "growth transformation" (GT).

Many times, the objective function of discriminative training of HMM can be formulated into a rational function, thus enabling the use of GT techniques. This type of techniques is also called extended Baum–Welch (EBW) algorithm when the underlying statistical model is an HMM. GT is an iterative optimization scheme where if the parameter set Λ is subject to a transformation $\Lambda = T(\Lambda')$, then the objective function "grows" in its value $O(\Lambda) > (\Lambda')$ unless $\Lambda = \Lambda'$. Hence the name growth transformation. GT or EBW algorithm was initially developed for the homogeneous polynomial by Baum and his colleagues (e.g., [4]). It was later extended to optimizing nonhomogeneous rational functions as reported in [14]. EBW algorithm became popular for its successful use in discriminative training of HMM using the maximum mutual information (MMI; see Chapter 3) criterion after the extension of the MMI training was made from the discrete HMM in [14] to the CD HMM in [3, 17, 34, 50, 52].

The importance of the optimization technique based on GT/EBW algorithm lies in its effectiveness and closed-form parameter updating for large-scale optimization problems with difficult objective functions (i.e., training criteria). With the traditional ML training where the likelihood function as the optimization criterion is relatively simple, a fast method is often available, such as the expectation–maximization (EM) algorithm for the HMM. In contrast, for the discriminative training criteria that are more complex than the ML, optimization becomes more difficult. For them, two general types of optimization techniques are available for the HMM: (1) gradient-based method and (2) GT/EBW. The latter has the advantage of having closed-form parameter updating formulas while not explicitly requiring second-order statistics. In addition, it does not require the same type of special and often delicate care for tuning the parameter-dependent learning rate as in the gradient-based methods (e.g., [25, 44]).

Let $G(\Lambda)$ and $H(\Lambda)$ be two real valued functions on the parameter set Λ, and the denominator function $H(\Lambda)$ is positive valued. And let the objective function be the ratio of them, giving the rational function of

$$O(\Lambda) = \frac{G(\Lambda)}{H(\Lambda)} \qquad (1.26)$$

a GT-based optimization algorithm exists to maximize $O(\Lambda)$.

An example of this rational function is the objective function for discriminative learning of HMM parameters, which will be discussed in greater details in Section 3.2.2, where

$$G(\Lambda) = \sum_s p(X, s|\Lambda) \; C(s) \text{ and } H(\Lambda) = \sum_s p(X, s|\Lambda) \qquad (1.27)$$

and we use $s = s_1, \ldots, s_R$ to denote a possible label sequences for all R training tokens, and use $X = x_1, \ldots, x_R$ to denote the observation data sequences for all R training tokens.

As originally proposed in [14], for the objective function of (1.26), the GT-based optimization algorithm constructs the auxiliary function of

$$F(\Lambda; \Lambda') = G(\Lambda) - O(\Lambda')H(\Lambda) + D \qquad (1.28)$$

where D is a quantity independent of the parameter set Λ, and Λ' denotes the parameter set obtained from the immediately previous iteration of the algorithm.

The algorithm starts by initializing the parameter set as, say, Λ'. (This is often accomplished by the ML training using, for instance, EM or Baum–Welch algorithm for HMMs.) Then, the updating of the parameter set from Λ' to Λ proceeds by maximizing the auxiliary function $F(\Lambda; \Lambda')$, and the process iterates until convergence is reached. Maximizing the auxiliary function $F(\Lambda; \Lambda')$ is often easier than maximizing the original rational function $O(\Lambda)$. And the following is a simple proof that as long as D is a quantity not relevant to the parameter set Λ, an increase of $F(\Lambda; \Lambda')$ guarantees an increase of $O(\Lambda)$.

Substituting $\Lambda = \Lambda'$ into (1.28), we have

$$F(\Lambda'; \Lambda') = \underbrace{G(\Lambda') - O(\Lambda')H(\Lambda')}_{=0} + D = D$$

Hence,

$$F(\Lambda; \Lambda') - F(\Lambda'; \Lambda') = F(\Lambda; \Lambda') - D = G(\Lambda) - O(\Lambda')H(\Lambda)$$
$$= H(\Lambda) \left(\frac{G(\Lambda)}{H(\Lambda)} - O(\Lambda') \right) = H(\Lambda) \left(O(\Lambda) - O(\Lambda') \right)$$

Because $H(\Lambda)$ is positive, we have $O(\Lambda) - O(\Lambda') > 0$ on the right-hand side if $F(\Lambda; \Lambda') - F(\Lambda; \Lambda') > 0$ on the left-hand side. That is, for optimizing a complicated rational function, we can turn the problem to optimizing $F(\Lambda; \Lambda')$, which is often simpler.

In later chapters of this book, we will provide details of optimizing $F(\Lambda; \Lambda')$ for discriminative training of speech recognizer parameters.

1.6 ORGANIZATION OF THE BOOK

The main content of this book is an extensive account of the discriminative learning techniques that are currently popular in training HMM-based speech recognition systems. In this introductory chapter, we first clarify the concepts of discriminative learning and speech recognition, and then we proceed to discuss the roles of discriminative learning in speech recognition practice. We then introduce several basic probability distributions that will be used in the remainder of this book and that also serve as the building blocks for the more complex distributions such as HMMs. Finally, we introduce some basic concepts and techniques of optimization including the definition of optima, a necessary condition for achieving the optima, Lagrange multiplier method, and gradient descent method. We also provide preliminaries to the growth-transformation based optimization technique that applies to specific, rational forms of the objective functions naturally fitting to those in discriminative learning of popular distributions such as HMMs used in speech recognition.

In Chapter 2, we will provide a tutorial on statistical speech recognition and on the state-of-the-art modeling techniques, setting up the context in which discriminative learning is motivated and applied to. In particular, HMMs are formally introduced. In Chapter 3, we provide a unified account for the several common objective functions for discriminative training of HMMs currently in use in speech recognition practice. We also compare our unified form of these objective functions with another form in literature. How to do discriminative parameter learning using the unified form of objective functions via the GT technique is discussed in Chapters 4 and 5; Chapter 4 deals with exponential family distribution parameters, and Chapter 5 focuses on more difficult HMM parameters. Some practical implementation issues of the GT technique for HMM parameter learning are discussed in Chapter 6. In Chapter 7, selected experimental results in speech recognition are presented. Finally, an epilogue and summary is given in Chapter 8.

·　　·　　·　　·　　·

CHAPTER 2

Statistical Speech Recognition: A Tutorial

In this chapter, we provide a tutorial on statistical speech recognition. In particular, we establish hidden Markov models (HMMs) as a principal modeling tool for characterizing acoustic features in speech. The purpose of this chapter is to set up the context in which HMM parameter learning and discriminative learning in particular, will be introduced.

2.1 INTRODUCTION

A key to understanding the human speech process is the dynamic characterization of its sequential or variable-length pattern. Current state-of-the-art speech recognition systems are mainly based on HMMs for acoustic modeling. In general, it is assumed that the speech signal and its features are a realization of some semantic or linguist message encoded as a sequence of linguistic symbols. To recognize the underlying symbol sequence given a spoken utterance, the speech waveform is first converted into a sequence of feature vectors equally spaced in time. Each feature vector is assumed to represent the speech waveform over a short duration of 10–30 ms, wherein the speech waveform is regarded as a stationary signal. Typical parametric representations include linear prediction coefficients, perceptual linear prediction, and Mel frequency cepstral coefficients, plus their time derivatives. Furthermore, these vectors are usually considered independent observations given a state of HMM.

As illustrated in Figure 2.1, the role of speech recognizer is to map a sequence of observation vectors into its underlying words. Let the speech signal be represented by a sequence of observation vectors X,

$$X = x_1, x_2 \ldots, x_T$$

where x_t is the speech vector observed at time t. The speech recognition problem can therefore be regarded as looking for the most possible word sequence S^* given the observation vector sequence X, that is,

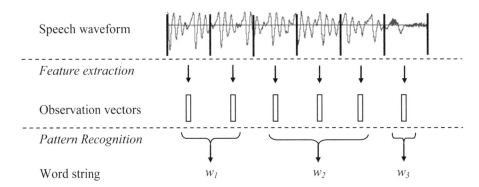

FIGURE 2.1: Illustration of the speech recognition process. The raw waveform of speech is first parameterized to discrete observation vectors. Then the word string that corresponds to that observation sequence is decoded by the recognizer.

$$S^* = \arg\max_s P(s|X) \tag{2.1}$$

According to Bayes rule, it is equivalent to solving S^* by:

$$S^* = \arg\max_s P(s, X) P(X) = \arg\max_s P(X|s) P(s) \tag{2.2}$$

where $P(s)$ is prior probability of the word sequence s, which is determined by a language model (LM), and $P(X|s)$ is the conditional probability or likelihood of X given s, which is computed from the acoustic model (AM) of the speech recognition system.

2.2 LANGUAGE MODELING

As described in the previous section, the a priori probability of the word sequence S is determined by the language model. For isolated-word speech recognition where recognition targets are isolated words. Given a K-word vocabulary, $P(w_i)$ is assigned to $1/K$ if a uniform distribution of word occurrence is assumed, or $P(w_i)$ can be determined by counting the occurrence frequency of word w_i in the language model training text corpus.

In continuous speech recognition, the computation of $P(S)$ is more complicated. Assume that the word sequence S has M words, that is,

$$S = w_1, w_2 \ldots, w_M$$

The probability of the word sequence S can be calculated as,

$$P(S) = P(w_1, w_2, \ldots, w_M) = P(w_1) \cdot \prod_{m=2}^{M} P(w_m | w_1, \ldots, w_{m-1}) \qquad (2.3)$$

Assuming that the word sequence is produced by a $(N-1)$th order Markov model, the computation of $P(S)$ can be simplified as

$$P(S) = P(w_1)P(w_2|w_1)\ldots P(w_{N-1}|w_1, \ldots, w_{N-2}) \cdot \prod_{m=N}^{M} P(w_m | w_{m-N+1}, \ldots, w_{m-1}) \quad (2.4)$$

This model is referred to as an N-gram language model.

Many papers have been published on how to reliably estimate an N-gram language model. The basic idea is to count the frequency of occurrences of each word in the LM training text corpus, given a particular word sequence that precedes the word. To handle possible word sequences that are not seen in the training text, a back-off mechanism is normally used to assign lower-bound scores to those rarely seen word sequences. In most speech recognition systems, bigram and trigram LMs are used.

2.3   ACOUSTIC MODELING AND HMMs

In speech recognition, statistical properties of sound events are described by the acoustic model. Correspondingly, the likelihood score $p(X|s)$ in Eq. (2.2) is computed based on the acoustic model. In HMM-based speech recognition, it is assumed that the sequence of observed vectors corresponding to each word is generated by a Markov chain. For large-vocabulary automatic speech recognition (LVASR), usually an HMM is constructed for each phone, then the HMM of a word is constructed by concatenating corresponding phone-specific HMMs. We can further concatenate HMMs of words to construct the HMM of the whole string that contains multiple words. Then $p(X|s)$ is computed through $p(X|\lambda_s)$, where λ_s is the HMM of the strings.

As shown in Figure 2.2, an HMM is a finite-state machine that changes state once every time frame, and at each time frame t when a state j is entered, an observation vector x_t is generated from the emitting probability distribution $b_j(x)$. The transition property from state i to state j is specified by the transition probability $a_{i,j}$. Moreover, two special nonemitting states are usually used in an HMM. They include an entry state, which is reached before the speech vector generation process begins, and an exit state, which is reached when the generative process terminates. Both states are reached only once. Because they do not generate any observation, none of them has an emitting probability density.

In the HMM, the transition probability $a_{i,j}$ is the probability of entering state j given the previous state i, that is, $a_{i,j} = \Pr(q_t = j \mid q_{t-1} = i)$, where q_t is the state index at time t. For an N-state HMM, we have,

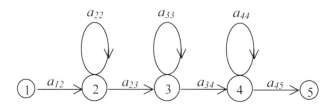

FIGURE 2.2: Illustration of a five-state left-to-right HMM. It has two nonemitting states and three emitting states. For each emitting state, the HMM is only allowed to remain at the same state or move to the next state.

$$\sum_{j=1}^{N} a_{i,j} = 1 \qquad (2.5)$$

The emitting probability density $b_j(x)$ describes the distribution of the observation vectors at state j. In discrete HMM (DHMM), emitting probability is represented by a multinomial distribution, whereas in continuous-density HMM (CDHMM), emitting probability density is often represented by a Gaussian mixture density:

$$b_j(x) = \sum_{m=1}^{M} c_{j,m} N\left(x; \mu_{j,m}, \Sigma_{j,m}\right) \qquad (2.6)$$

where $N(x; \mu_{j,m}, \Sigma_{j,m}) = \dfrac{1}{(2\pi)^{\frac{D}{2}} |\Sigma_{j,m}|^{\frac{1}{2}}} e^{-\frac{1}{2}(x-\mu_{j,m})^T \Sigma_{j,m}^{-1}(x-\mu_{j,m})}$ is a multivariate Gaussian density, D is the dimension of the feature vector x, and cjm, mjm, and Sjm are the weight, mean, and covariance of the mth Gaussian component of the mixture distribution at state j. Generally speaking, each emitting distribution characterizes a sound event, and the distribution must be specific enough to allow discrimination between different sounds as well as robust enough to account for the variability in natural speech.

Given $\{a_{i,j}\}$ and $b_j(x)$, for $i = 1, 2, \ldots, N, j = 1, 2, \ldots, N$, the likelihood of an observation sequence X is calculated as:

$$p(X|\lambda) = \sum_{q} p\left(X, q|\lambda\right) \qquad (2.7)$$

where $q = q_1, q_2, \ldots, q_T$ is the HMM state sequence that generates the observation vector sequence $X = x_1, x_2, \ldots, x_T$, and the joint probability of X and the state sequence q given λ is a product of the transition probabilities and the emitting probabilities

$$p(X, q|\lambda) = \prod_{t=1}^{t} b_{q_t}(x_t) a_{q_{t-1}, q_t} \qquad (2.8)$$

where q_{T+1} is the nonemitting exit state.

In practice, (2.7) can be approximately calculated as joint probability of the observation vector sequence X with the most possible state sequence, that is,

$$p(X|\lambda) \approx \max_{q} p(X, q|\lambda) \qquad (2.9)$$

Although it is impractical to evaluate the quantities of (2.7) and (2.9) directly due to the huge number of possible state sequences when T is large, efficient recursive algorithms such as forward–backward method and Viterbi decoding method exist for computing them [10, 43].

CHAPTER 3

Discriminative Learning: A Unified Objective Function

In this chapter, a unified account is provided for three classes of objective functions developed in discriminative training of hidden Markov models (HMMs). These are: maximum mutual information (MMI), minimum classification error (MCE), and minimum phone error/minimum word error (MPE/MWE). We also compare our unified form of these objective functions with another popular unified form in the literature.

3.1   INTRODUCTION

Popular discriminative parameter learning techniques are (1) MMI [6, 14, 17, 34, 49, 52]; (2) MCE [8, 20, 24, 25, 31–33, 42, 44, 46], and (3) MPE and closely related MWE [12, 38–41]. In addition to a general overview on the above classes of techniques, this book has a special focus on three key areas in discriminative learning: objective function, optimization method, and algorithmic properties. This chapter is devoted to the first area, where we provide a unified view of the three discriminative learning objective functions, MMI, MCE, and MPE/MWE, for classifier parameter optimization, from which structural insight and the relationships among them are derived. In this chapter, we concentrate on a unified objective function that gives rise to various special cases associated with different levels of performance optimization for pattern recognition tasks — including the performance optimization levels of superstring unit, string unit, and substring unit.

After giving an introduction to the discriminative learning criteria of MMI, MCE, and MPE/MWE, we show that under certain assumptions, the objective functions of MMI, MCE, and MPE/MWE criteria (with multiple training tokens) can be formulated and unified into a rational-function form. From that, relations among MMI, MCE, and MPE/MWE criteria are studied. In discussing these topics, some familiarities of HMMs are assumed, such as those described in standard textbooks (e.g., [43, 47]).

3.2   A UNIFIED DISCRIMINATIVE TRAINING CRITERION

The main purpose of this chapter is to provide a general and concise introduction to three types of optimization criteria, MMI, MCE, and MPE/MWE, for discriminative parameter learning, and then to formulate a unified criterion that subsumes the three criteria as special cases. The process of this formulation offers insight into the fundamental relationship among MMI, MCE, and MPE. Another insight gained is on how these special cases correspond to distinct levels of pattern recognition performance optimization. MMI gives performance optimization for superstring sequences. MCE gives performance optimization for string sequences. And MPE/MWE gives performance optimization for substring sequences.

3.2.1   Notations

As the notations throughout this book, we denote by Λ the parameter set of the generative model (e.g., HMM or a Gaussian distribution) expressed in terms of a joint statistical distribution:

$$p(X, S | \Lambda) = p(X | S, \Lambda) P(S) \qquad (3.1)$$

on the observation training data sequence X and on the corresponding label sequence S, where we assume the parameters in the "language model" $P(S)$ are not subject to optimization. We denote by R the number of training tokens and use $r = 1, \ldots, R$ as the index for "token" or "string" (e.g., a single sentence or utterance) in the training data, and each token consists of a "string" of an observation data sequence: $X_r = x_{r,1}, \ldots, x_{r,Tr}$ of length T_r with the corresponding label (e.g., word) sequence: $S_r = w_{r,1}, \ldots, w_{r,Nr}$ of length N_r. That is, S_r denotes the correct label sequence for token r; in effect, $w_{r,i}$ is the ith word in the word sequence of S_r. Furthermore, we use s_r to denote all possible label sequences for the rth token, including the correct label sequence S_r and all other incorrect label sequences. For the iterative learning approach discussed in this book, we denote by Λ' the model parameters computed from the immediately previous iteration.

3.2.2   The Central Result

The central result presented in this chapter is that all three discriminative learning criteria, MMI, MCE, and MPE/MWE, can be formulated as the following unified form of a rational function as the objective function (which can be readily subject to a special way of optimization discussed later):

$$O(\Lambda) = \frac{\sum_{s_1, \ldots, s_R} p(X_1, \ldots, X_R, s_1, \ldots, s_R | \Lambda) \cdot C_{\text{DT}}(s_1, \ldots, s_R)}{\sum_{s_1, \ldots, s_R} p(X_1, \ldots, X_R, s_1, \ldots, s_R | \Lambda)} \qquad (3.2)$$

where the summation over $s = s_1, \ldots, s_R$ in (3.2) denotes the coverage of all possible label sequences (both correct and incorrect ones) in all R training tokens. (This huge number of terms will be drastically simplified during the optimization step, which we shall discuss in detail later.) In (3.2), $X = X_1, \ldots, X_R$ denotes the collection of all observation data sequences (strings) in all R training tokens, which we also call "superstring." $p_\Lambda(X_1, \ldots X_R, s_1, \ldots, s_R)$ is the joint distribution for the super-string of data $X_1, \ldots, X_R$ and an arbitrary possible super-string label sequence assignments $s_1, \ldots, s_R$. The discriminative training (DT) function $C_{DT}(s_1 \ldots, s_R)$ in (3.2) differentiates MMI, MCE, and MPE/MWE, each with a specific form of $C_{MMI}(s_1 \ldots, s_R)$, $C_{MCE}(s_1 \ldots, s_R)$, and $C_{MPE}(s_1 \ldots, s_R)$, respectively. We will derive these specific forms in the subsequent sections of this chapter. Note that $C_{DT}(s_1 \ldots, s_R)$ in (3.2) is a quantity that depends only on the label sequence $s_1, \ldots, s_R$, and are independent of the parameter set Λ to be optimized.

We now introduce the criteria of MMI, MCE, and MPE/MWE separately as in the standard literature, and then provide detailed derivations to reformulate each of them into the unified rational function form of (3.2). This then will enable the use of powerful and unified optimization techniques based on GT applied specifically to rational functions.

3.3   MMI AND ITS UNIFIED FORM

3.3.1   Introduction to MMI Criterion

In information theory, mutual information $I(X,S)$ between data X and their corresponding labels/symbols S measures the amount of information obtained, or the amount of reduction in uncertainty, through a noisy information-transfer channel after observing output labels/symbols. In designing the noisy channel, it is obvious that one desires to increase the information attainment by maximizing $I(X,S)$. Quantitatively, mutual information is defined as

$$I(X, S) = \sum_{X,S} p(X, S) \log \frac{p(X, S)}{p(X)p(S)} = \sum_{X,S} p(X, S) \log \frac{p(S|X)}{p(S)} = H(S) - H(S|X) \qquad (3.3)$$

where $H(S) = -\sum_s p(S) \log p(S)$ is the entropy of S, and $H(S|X)$ is the conditional entropy:

$$H(S|X) = -\sum_{X,S} p(X, S) \log P(S|X) \qquad (3.4)$$

Assume that $P(S)$ ("language model") and hence $H(S)$ is given (i.e., with no parameters to optimize). Then maximizing mutual information of (3.3) becomes equivalent to minimizing conditional entropy of (3.4) with respect to its parameters. Because $P(S|X)$ in (3.4) is unknown, $H(S|X)$ can only be estimated using the sample average:

$$H(S|X) \cong \hat{H}_\Lambda(S|X) = -\frac{1}{R} \sum_{r=1}^{R} \log p(S_r|X_r) = -\frac{1}{R} \sum_{r=1}^{R} \log \frac{p(X_r|S_r, \Lambda)P(S_r)}{p(X_r|\Lambda)}$$

Hence, maximizing mutual information (MMI) is equivalent to maximizing

$$O_{\text{MMI}}(\Lambda) = \sum_{r=1}^{R} \log \frac{p(X_r|S_r,\Lambda)P(S_r)}{p(X_r|\Lambda)} = \sum_{r=1}^{R} \log \frac{p(X_r|S_r,\Lambda)P(S_r)}{\sum_{s_r} p(X_r|s_r,\Lambda)P(s_r)} \qquad (3.5)$$

where $P(s_r)$ is the "language model" probability for an arbitrary sentence token s_r. The MMI crite-rion equals the logarithm of the posterior probability of the correct sentence S_r, or "good model," given their observation sequences. This posterior probability takes into account all models, good (S_r) or bad (s_r excluding S_r), as shown in the denominator of (3.5). (In practice, a scale κ has been applied empirically to all the probability terms in (3.5) for generalization purposes in implementing MMI [40]. This issue will not be addressed in this book).

3.3.2 Reformulation of the MMI Criterion into Its Unified Form

It is straightforward to reformulate the problem of optimizing (3.5) into that of optimizing the uni-fied form of (3.2), because (3.5) is essentially a rational function due to the logarithm. To continue the reformulation, we construct the monotonically increasing function of exponentiation for (3.5). This gives

$$\tilde{O}_{\text{MMI}}(\Lambda) = \exp[O_{\text{MMI}}(\Lambda)] = \prod_{r=1}^{R} \frac{p(X_r, S_r|\Lambda)}{\sum_{s_r} p(X_r, s_r|\Lambda)} = \frac{p(X_1, \ldots, X_R, S_1, \ldots, S_R|\Lambda)}{\sum_{s_1,\ldots,s_R} p(X_1,\ldots,X_R,s_1,\ldots,s_R|\Lambda)} \qquad (3.6)$$

The latter step uses the assumption that different training tokens are independent of each other. It is noteworthy that each multiplier in (3.6) can be viewed as a model-based expected utility, that is,

$$\frac{p(X_r,S_r|\Lambda)}{\sum_{s_r} p(X_r,s_r|\Lambda)} = 1 - \sum_{s_r \neq S_r} P(s_r|X_r,\Lambda) = 1 - \overbrace{\sum_{s_r} \underbrace{(1 - \delta(s_r,S_r))}_{0-1 \text{ loss}} P(s_r|X_r,\Lambda)}^{\Lambda\text{-based expected loss}}$$

We now rewrite (3.6) in the form of a rational function

$$\tilde{O}_{\text{MMI}}(\Lambda) = \frac{\sum_{s_1,\ldots,s_R} p(X_1,\ldots,X_R,s_1,\ldots,s_R|\Lambda)\, C_{\text{MMI}}(s_1,\ldots,s_R)}{\sum_{s_1,\ldots,s_R} p(X_1,\ldots,X_R,s_1,\ldots,s_R|\Lambda)} \qquad (3.7)$$

where

$$C_{\text{MMI}}(s_1,\ldots,s_R) = \prod_{r=1}^{R} \delta(s_r, S_r) \qquad (3.8)$$

is a quantity that depends only on the sentence sequence $s_1, \ldots, s_R$. In (3.8), $\delta(s_r, S_r)$ is the Kronecker delta function, that is, $\delta(s_r, S_r) = \begin{cases} 1 \text{ if } s_r = S_r \\ 0 \text{ otherwise} \end{cases}$.

We note that MMI is a discriminative performance measure at the "superstring" level in that it aims to improve the conditional likelihood on the entire training data set instead of on each individual string (token). This is reflected by the product form of the function in (3.8). $C_{\mathrm{MMI}}(s_1, \ldots, s_R)$ in (3.8) can be interpreted as the binary function (as "accuracy count") of the "superstring" $s_1, \ldots, s_R$, which takes a value of 1 if the superstring $s_1, \ldots, s_R$ is correct and zero otherwise. Correspondingly, $O_{\mathrm{MMI}}(\Lambda)$ can be interpreted as the average superstring accuracy count of the full training data set, which takes a continuous value between 0 and 1.

3.4   MCE AND ITS UNIFIED FORM

The key result of this section is to reformulate another popular discriminative criterion, that is, MCE, into the same form of the rational function as in (3.7), except that the accuracy-count function $C(\cdot)$ takes a summation form instead of a product form. This correspondingly gives the string-level discriminative performance measure for MCE, contrasting with the superstring level for the MMI criterion just described. We now first provide a concise introduction to the basic concept and conventional formulation of MCE.

3.4.1   Introduction to the MCE Criterion

MCE learning was originally introduced for multiple-category classification problems where the smoothed error rate is minimized for isolated "tokens" [2, 24]. It was later generalized to minimize the smoothed "sentence token" or string-level error rate [8, 25], which is known as "embedded MCE." The MCE objective function is defined first based on a set of discriminant functions and a special type of loss function. Then model parameters are estimated to minimize the expected loss that is closely related to the recognition error rate of the classifier.

In embedded MCE training, for the rth training token, a set of discriminant functions is defined as the log likelihood of data based on the correct as well as competing strings:

$$g_{s_r}(X_r; \Lambda) = \log p(X_r, s_r | \Lambda) \tag{3.9}$$

Then the decision rule of the classifier/recognizer can be expressed as

$$C(X_r) = \overset{*}{s_r} \text{ iff } \overset{*}{s_r} = \arg\max_{s_r} g_{s_r}(X_r; \Lambda) \tag{3.10}$$

For sequential pattern recognition tasks such as continuous speech recognition, usually only the N most confusable competing strings, $s_{r,1}, \ldots, s_{r,N}$, are considered in MCE. Note these N

confusable competing strings change dynamically after each training iteration. In practice, they are regenerated after each iteration through N-best decoding based on the parameters Λ' obtained at the immediately previous iteration. The N-best strings can be defined inductively by

$$s_{r,1} = \underset{s_r:s_r \neq S_r}{\arg\max} \log p(X_r, s_r|\Lambda) \approx \underset{s_r:s_r \neq S_r}{\arg\max} \log p(X_r, s_r|\Lambda')$$

$$s_{r,i} = \underset{s_r:s_r \neq S_r, s_r \neq s_{r,1}, \dots, s_{r,i-1}}{\arg\max} \log p(X_r, s_r|\Lambda) \approx \underset{s_r:s_r \neq S_r, s_r \neq s_{r,1}, \dots, s_{r,i-1}}{\arg\max} \log p(X_r, s_r|\Lambda') \quad i = 2, \dots, N.$$

$$(3.11)$$

Next, a misclassification measure $d_r(X_r, \Lambda)$ is defined to emulate the decision rule for utterance r, that is, $d_r(X_r, \Lambda) \geq 0$ implies misclassification and $d_r(X_r, \Lambda) < 0$ implies correct classification,

$$d_r(X_r, \Lambda) = -g_{S_r}(X_r; \Lambda) + G_{S_r}(X_r; \Lambda) \qquad (3.12)$$

where $G_{S_r}(X_r; \Lambda)$ is a function that represents the score of incorrect strings competing with the correct string S_r.

For 1-best MCE training ($N = 1$), only the best-incorrect-string $s_{r,1}$ is considered as the competitor. In this special case, $G_{S_r}(X_r; \Lambda)$ clearly becomes

$$G_{S_r}(X_r; \Lambda) = g_{s_{r,1}}(X_r; \Lambda) \qquad (3.13)$$

However, for the general case where $N > 1$, different definitions of $G_{S_r}(X_r; \Lambda)$ can be used. One popular definition takes the following form [25]:

$$G_{S_r}(X_r; \Lambda) = \log \left\{ \frac{1}{N} \sum_{i=1}^{N} p^\eta(X_r, s_{r,i}|\Lambda) \right\}^{\frac{1}{\eta}} \qquad (3.14)$$

Another popular form of $g_{S_r}(X_r; \Lambda)$ and $G_{S_r}(X_r; \Lambda)$ (the latter has similar effects to (3.14) and was used in [46]) is

$$\begin{cases} g_{S_r}(X_r; \Lambda) = \log p^\eta(X_r, S_r|\Lambda) \\ G_{S_r}(X_r; \Lambda) = \log \sum_{i=1}^{N} p^\eta(X_r, s_{r,i}|\Lambda) \end{cases} \qquad (3.15)$$

where η is a scaling factor for joint probability $p(X_r, s_r|\Lambda)$. In this paper, we adopt $G_{S_r}(X_r; \Lambda)$ with the form of (3.15) and set $\eta = 1$ for simplicity and mathematic tractability. (We will discuss the $\eta \neq 1$ case in Chapter 6.)

Now we define the MCE loss function, which, for a single utterance r, is typically a sigmoid function as originally proposed in [24, 25]:

$$l_r(d_r(X_r, \Lambda)) = \frac{1}{1 + e^{-\alpha d_r(X_r, \Lambda)}} \qquad (3.16)$$

where α is the slope of the sigmoid function, often determined empirically. As presented in [21] (p. 156), we also use $\alpha = 1$ for simplicity in exposition. In practice, however, α is usually set to be a value less than 1; we will discuss this more general case in Chapter 6. Note that the loss function of (3.16) emulates the desirable zero–one classification error count.

Using the misclassification measure in the form of (3.12) and (3.15) (with $\eta = 1$), we can rewrite the loss function for one training string token as

$$l_r(d_r(X_r, \Lambda)) = \frac{\sum\limits_{s_r, s_r \neq S_r} p(X_r, s_r | \Lambda)}{\sum\limits_{s_r, s_r \neq S_r} p(X_r, s_r | \Lambda) + p(X_r, S_r | \Lambda)} = \frac{\sum\limits_{s_r, s_r \neq S_r} p(X_r, s_r | \Lambda)}{\sum\limits_{s_r} p(X_r, s_r | \Lambda)} \qquad (3.17)$$

which can be viewed as an model-based expected loss of classifying X_r to S_r, after putting it in the following form:

$$l_r(d_r(X_r, \Lambda)) = \sum\limits_{s_r \neq S_r} P(s_r | X_r, \Lambda) = \sum\limits_{s_r} \underbrace{(1 - \delta(s_r, S_r))}_{0-1 \text{ loss}} P(s_r | X_r, \Lambda)$$

Then, because the error count sums over training tokens, the loss function for all R training tokens is naturally defined to be:

$$L_{\text{MCE}}(\Lambda) = \sum\limits_{r=1}^{R} l_r(d_r(X_r, \Lambda)) \qquad (3.18)$$

Here, we emphasize the summation in (3.18) for combining all string tokens for MCE. Because each loss function approximates the string error count, the total empirical error count rightfully becomes the sum of all independent individual string error counts. This forms a sharp contrast to the MMI case as in (3.6), where a product of probabilities is constructed in pooling all string tokens.

Now, minimizing the overall loss function of $L_{\text{MCE}}(\Lambda)$ in (3.18) is equivalent to maximizing the following MCE objective function:

$$O_{\text{MCE}}(\Lambda) = R - L_{\text{MCE}}(\Lambda) = \sum\limits_{r=1}^{R} \left[1 - \frac{\sum\limits_{s_r, s_r \neq S_r} p(X_r, s_r | \Lambda)}{\sum\limits_{s_r} p(X_r, s_r | \Lambda)} \right] = \sum\limits_{r=1}^{R} \frac{p(X_r, S_r | \Lambda)}{\sum\limits_{s_r} p(X_r, s_r | \Lambda)} \qquad (3.19)$$

3.4.2   Reformulation of the MCE Criterion Into its Unified Form

Unlike the MMI case, the MCE objective function as expressed in (3.19) is a sum of rational functions rather than a rational function in itself, and hence it would not be amenable to the special form of GT-based optimization. The state-of-the-art techniques for optimizing the MCE objective function have been based on gradient descent, which is called generalized probabilistic descent (GPD) [8, 24, 25]. As one original contribution of this paper, we reformulate the MCE objective function as a true rational function in a nontrivial fashion. This not only unifies the earlier disparate types of objective functions and offers insights into their differences and similarities, but, more importantly, it enables the use of GT as an alternative technique to GPD for faster and more effective optimization.

The reformulation proceeds as follows:

$$O_{\text{MCE}}(\Lambda) = \sum_{r=1}^{R} \frac{\sum_{s_r} p(X_r, s_r | \Lambda) \delta(s_r, S_r)}{\sum_{c} p(X_r, s_r | \Lambda)} \tag{3.20}$$

$$= \underbrace{\frac{\sum_{s_1} p(X_1, s_1 | \Lambda) \delta(s_1, S_1)}{\sum_{s_1} p(X_1, s_1 | \Lambda)}}_{:=O_1} + \underbrace{\frac{\sum_{s_2} p(X_2, s_2 | \Lambda) \delta(s_2, S_2)}{\sum_{s_2} p(X_2, s_2 | \Lambda)}}_{:=O_2}$$

$$+ \underbrace{\frac{\sum_{s_3} p(X_3, s_3 | \Lambda) \delta(s_3, S_3)}{\sum_{s_3} p(X_3, s_3 | \Lambda)}}_{:=O_3} + \cdots + \underbrace{\frac{\sum_{s_R} p(X_R, s_R | \Lambda) \delta(s_R, S_R)}{\sum_{s_R} p(X_R, s_R | \Lambda)}}_{:=O_R}$$

$$= \frac{\sum_{s_1} \sum_{s_2} p(X_1, s_1 | \Lambda) p(X_2, s_2 | \Lambda) [\delta(s_1, S_1) + \delta(s_2, S_2)]}{\sum_{s_1} \sum_{s_2} p(X_1, s_1 | \Lambda) p(X_2, s_2 | \Lambda)} + O_3 + \cdots + O_R$$

$$= \frac{\sum_{s_1 s_2} p(X_1, X_2, s_1, s_2 | \Lambda) [C_{\text{MCE}}(s_1 s_2)]}{\sum_{s_1 s_2} p(X_1, X_2, s_1, s_2 | \Lambda)} + O_3 + \cdots + O_R$$

$$= \frac{\sum_{s_1 s_2 s_3} p(X_1, X_2, X_3, s_1, s_2, s_3 | \Lambda) [C_{\text{MCE}}(s_1 s_2 s_3)]}{\sum_{s_1 s_2 s_3} p(X_1, X_2, X_3, s_1, s_2, s_3 | \Lambda)} + \cdots + O_R$$

$$= \frac{\sum_{s_1,\ldots,s_R} p(X_1,\ldots,X_R,s_1,\ldots,s_R|\Lambda)\, C_{\mathrm{MCE}}(s_1,\ldots,s_R)}{\sum_{s_1,\ldots,s_R} p(X_1,\ldots,X_R,s_1,\ldots,s_R|\Lambda)} \qquad (3.21)$$

where $C_{\mathrm{MCE}}(s_1,\ldots,s_R) = \sum_{r=1}^{R} \delta(s_1,S_R)$. The final result of (3.21) gives the rational function fitting exactly to the unified form of (3.2). The correctness of (3.21) can also be proved directly by induction, which we leave to readers as an exercise. $C_{\mathrm{MCE}}(s_1,\ldots,s_R)$ in (3.21) can be interpreted as the string accuracy count for $s_1,\ldots,s_R$, which takes an integer value between 0 and R as the number of correct strings in $s_1,\ldots,s_R$. Correspondingly, $O_{\mathrm{MCE}}(\Lambda)$ can be interpreted as the average string accuracy count of the full training data set.

3.5   MPE/MWE AND ITS UNIFIED FORM

3.5.1   Introduction to the MPE/MWE Criterion

In this section, we introduce yet another commonly used discriminative training objective function, MPE or MWE, in speech recognition, developed originally in [38, 40]. In contrast to MMI and MCE described earlier, which are aimed at the superstring level and at the string level of recognition performance optimization, respectively, MPE/MWE is aimed for performance optimization at the substring level. In speech recognition, a string corresponds to a sentence, and a substring as a constituent of the sentence can be words or phones. Because performance measures of speech recognition are often the word or phone error rates rather than the sentence error rate, it has been argued that MPE/MWE is a more appropriate criterion to optimize than the MMI and MCE criteria [40].

The MPE objective function that needs to be maximized is defined as

$$O_{\mathrm{MPE}}(\Lambda) = \sum_{r=1}^{R} \frac{\sum_{s_r} p(X_r,s_r|\Lambda) \sum_{s_r} A(s_r,S_r)}{\sum_{s_r} p(X_r,s_r|\Lambda)} \qquad (3.22)$$

where $A(s_r,S_r)$ is the raw phone (substring) accuracy count in the sentence string s_r (proposed originally in [38, 40]). Specifically, $A(s_r,S_r)$ is the total phone (substring) count in the reference string S_r minus the sum of insertion, deletion, and substitution errors of s_r computed based on S_r.

The MPE criterion (3.22) equals the model-based expectation of the raw phone accuracy count over the entire training set. This becomes clear by rewriting (3.22) as

$$O_{\mathrm{MPE}}(\Lambda) = \sum_{r=1}^{R} \sum_{s_r} p(s_r|X_r,\Lambda) A(s_r,S_r)$$

where

$$p(s_r|X_r,\Lambda) = \frac{p(X_r,s_r|\Lambda)}{p(X_r|\Lambda)} = \frac{p(X_r,s_r|\Lambda)}{\sum_{s_r} p(X_r,s_r|\Lambda)}$$

is the model-based posterior probability over which the expectation is taken in defining the MPE objective function of (3.22). It can be shown that MPE criterion provides an upper bound of true Bayes risk on the substring (e.g., phone) level.

When $A(s_r, S_r)$ in (3.22) is changed from the raw phone accuracy count to another raw substring accuracy for words $A_l(s_r, S_r)$, we have the virtually equivalent definition of the MWE criterion:

$$O_{\mathrm{MWE}}(\Lambda) = \sum_{r=1}^{R} \frac{\sum_{s_r} p(X_r|s_r,\Lambda)\ P(s_r)A_l(s_r, S_r)}{\sum_{s_r} p(X_r|s_r,\Lambda)\ P(s_r)} \qquad (3.23)$$

and hence, throughout this book, we merge these two into the same MPE/MWE category.

3.5.2 Reformulation of the MPE/MWE Criterion Into Its Unified Form

The MPE/MWE objective function is also a sum of multiple rational functions instead of a single rational function, and hence it is difficult to derive GT formulas, as pointed out in [40] (Section 7.2, p. 92). The state-of-the-art techniques for optimizing the MPE/MWE objective functions have been based on a weak-sense auxiliary function (WSAF) proposed in [40], where the difficulty of formulating a rational function and the desire of moving away from traditional gradient descent have been eloquently discussed. In this paper, we propose to reformulate the MPE/MWE objective functions as a unified rational function in the form of (3.2). This enables an alternative technique to WSAF for optimization but with guaranteed convergence in the algorithm's iteration. The re-formulation of the MPE/MWE criteria (3.22)–(3.23) in the unified form of rational function follows the same steps as in the preceding MCE case. Note that (3.22)–(3.23) are in the same form as (3.20), except for the replacement of $\delta(s_r, S_r)$ by $A(s_r, S_r)$ or $A_1(s_r, S_r)$. Then, the same steps starting from (3.20) to (3.21) lead to the reformulated results for MPE/MWE:

$$O_{\mathrm{MPE}}(\Lambda) = \frac{\sum_{s_1,\ldots,s_R} p(X_1,\ldots,X_R,s_1,\ldots,s_R|\Lambda)\,C_{\mathrm{MPE}}(s_1,\ldots,s_R)}{\sum_{s_1,\ldots,s_R} p(X_1,\ldots,X_R,s_1,\ldots,s_R|\Lambda)} \qquad (3.24)$$

where

$$C_{\text{MPE}}(s_1,\ldots,s_R) = \sum_{r=1}^{R} A(s_r, S_r)$$

and

$$O_{\text{MWE}}(\Lambda) = \frac{\displaystyle\sum_{s_1,\ldots,s_R} p(X_1,\ldots,X_R,s_1,\ldots,s_R|\Lambda) C_{\text{MWE}}(s_1,\ldots,s_R)}{\displaystyle\sum_{s_1,\ldots,s_R} p(X_1,\ldots,X_R,s_1,\ldots,s_R|\Lambda)} \qquad (3.25)$$

where

$$C_{\text{MWE}}(s_1,\ldots,s_R) = \sum_{r=1}^{R} A_l(s_r, S_r)$$

Note that $C_{\text{MPE}}(s_1, \ldots, s_R)$ in (3.24) or $C_{\text{MWE}}(s_1, \ldots, s_R)$ in (3.25) can be interpreted as the raw phone or word (substring unit) accuracy count within the "superstring" $s_1, \ldots, s_R$. Its upper-limit value is the total number of phones or words in the full training data (i.e., the correct superstring $S_1, \ldots, S_R$). The actual value may be negative if many insertion errors occur. Correspondingly, $O_{\text{MPE}}(\Lambda)$ and $O_{\text{MWE}}(\Lambda)$ can be interpreted as the average raw phone or word accuracy count of the full training data set.

3.6 DISCUSSIONS AND COMPARISONS

3.6.1 Discussion and Elaboration on the Unified Form

We first provide a summary of the previous sections in this chapter, where a rational-function form of the discriminative training (DT) objective function or criterion is established as in (3.2) that unifies MMI, MCE, and MPE/MWE. In this unified form, the choice of the set of label sequences and the form of the generic function $C_{\text{DT}}(s_1, \ldots, s_R)$ determine the particular DT criterion, as summarized in Table 3.1. As an example shown in Table 3.1, for MMI, we have the specific function $C_{\text{DT}}(s_1, \ldots, s_R) = \prod_{r=1}^{R} \delta(s_r, S_R)$. For MPE, the function becomes $C_{\text{DT}}(s_1, \ldots, s_R) = \sum_{r=1}^{R} A(s_r, S_r)$. For MCE with general N-best competitors where $N > 1$, $C_{\text{DT}}(s_1, \ldots, s_R) = \sum_{r=1}^{R} \delta(s_r, S_r)$. For 1-best MCE ($N = 1$), s_r belongs to only the subset $\{S_r, s_{r,1}\}$. Equation (3.2) allows direct comparisons of the MMI, MCE, and MPE/MWE criteria. The most important insight offered by the unified framework of (3.2) is that the difference of these three types of criteria is embedded only in the weighting of alternative strings, where the weights (i.e., $C_{\text{DT}}(s_1, \ldots, s_R)$) are independent of the model parameters Λ to be learned.

TABLE 3.1: A unified rational-function form of the DT objective function (3.2), where differences in $C_{DT}(s_1, ..., s_R)$ distinguish MMI, MCE, and MPE/MWE and the number of "competing token candidates" distinguishes N-best and 1-best versions of the MCE

OBJECTIVE FUNCTIONS	$C_{DT}(S_R)$	$C_{DT}(S_1,..., S_R)$	LABEL SEQUENCE SET USED IN DT
MCE (*N*-best)	$\delta(s_r, S_r)$	$\sum_{r=1}^{R} C_{DT}(s_r)$	$\{S_r, s_{r,1}, ..., s_{r,N}\}$
MCE (1-best)	$\delta(s_r, S_r)$	$\sum_{r=1}^{R} C_{DT}(s_r)$	$\{S_r, s_{r,1}\}$
MPE	$A(s_r, S_r)$	$\sum_{r=1}^{R} C_{DT}(s_r)$	all possible label sequences
MWE	$A_l(s_r, S_r)$	$\sum_{r=1}^{R} C_{DT}(s_r)$	all possible label sequences
MMI	$\delta(s_r, S_r)$	$\prod_{r=1}^{R} C_{DT}(s_r)$	all possible label sequences

Note that the overall $C_{DT}(s_1, ..., s_R)$ is constructed from its constituents $C_{DT}(s_r)$'s in individual string tokens by either summation (for MCE, MPE/MWE) or product (for MMI).

As pointed out in [40], MPE/MWE has an important difference from MCE and MMI in that the weighting given by the MPE/MWE criteria to an incorrect string (sentence token) depends on the number of wrong substrings (wrong phones or words) within the string. MCE and MMI make a binary distinction based on whether the entire sentence string is correct or not, which is not desirable when reduction of substring errors (e.g., word errors in speech recognition) is the main goal of the sequential pattern recognition tasks. This distinction is most clearly seen by the sum of the binary function $C_{DT}(s_1, ..., s_R) = \sum_{r=1}^{R} \delta(s_r, S_r)$ for MCE and the sum of nonbinary functions $C_{DT}(s_1, ..., s_R) = \sum_{r=1}^{R} A(s_r, S_r)$ for MPE/MWE, both within the same unified framework. This key difference gives rise to the distinction of the substring level versus the string level of recognition performance optimization associated with MPE/MWE and MCE, respectively. As it performs the sentence or string-level optimization, MCE tends to push and pack errors into a few sentence tokens so as to create as many error-free token "strings" as possible. It sacrifices word/phone (substring) errors in order to reduce string errors, which may not be desirable when high word or phone accuracy is the goal of continuous speech or phonetic recognition.

Furthermore, the product instead of the summation form of the binary function associated with MMI, that is, $C_{DT}(s_1, \ldots, s_R) = \prod_{r=1}^{R} \delta(s_r, S_R)$, makes it clear that MMI achieves performance optimization at the superstring level. That is, as long as any one single sentence token has an error, the product of the Kronecker delta functions becomes zero. Therefore, all the summation terms in the numerator of (3.2) are zero except for the one corresponding to the correct label/transcription sequence. This "superstring" level performance measure is apparently less desirable than MCE or MPE/MWE, as has been shown extensively in experiments [31, 38–40].

Another insight gleaned from the unified form of the objective function (3.2) is that in the case of having only one sentence token (i.e., $R = 1$) in the training data and when the sentence contains only one phone, then all three MMI, MCE, and MPE/MWE criteria become identical. This is because in this case $C_{DT}(s_1, \ldots, s_R)$ becomes identical for all these criteria. The difference surfaces only when the training set consists of multiple sentence tokens. In this realistic case, the difference lies only in $C_{DT}(s_1, \ldots, s_R)$ as the Λ-independent weighing factor (as well as in the set of competitor strings), whereas the general rational-function form for the three criteria remains unchanged.

The major benefit of unifying the discriminative training criteria into a single rational function as in (3.2) is that we can then extend the same, well-established framework of GT to optimize all these major discriminative training criteria. Moreover, it also provides a new possibility of applying other, more advanced rational function optimization methods to the various discriminative training criteria. As an example, Jebara [22, 23] proposed a novel optimization method for rational functions as an alternative to the traditional GT method. In [22], the reverse Jensen's inequality method was developed and described, based on which an elegant solution for rational function optimization for HMMs with exponential-family densities was constructed. We will review this method in Appendix V, showing that our unified framework of (3.2) is directly subject to the application of this method and that this is not the case for another framework that we review below.

3.6.2 Comparisons With Another Unifying Framework

In recent papers [31, 46], an approach was proposed to unify a number of discriminative learning methods including MMI, MPE, and MPE/MWE (the earlier paper [46] did not include MPE/MWE). Functional similarities and differences among MMI, MCE, and MPE/MWE criteria were noted and discussed in [31, 46]. The framework proposed in this paper takes an additional step of unifying these criteria in a common rational-function form, and GT-based discriminative

TABLE 3.2: Choices of the smoothing function $f(z)$, alternative word sequences M_r, and exponent weight η in (3.26) for various types of DT criteria

CRITERIA	SMOOTHING FUNCTION $f(z)$	ALTERNATIVE WORD SEQUENCES M_r	η
MCE (N-best)	$-1/[1 + \exp(2qz)]$	$\{s_r\}$ excluding S_r	≥ 1
MCE (1-best)	$-1/[1 + \exp(2qz)]$	$\{s_{r,1}\}$	N/A
MPE/MWE	$\exp(z)$	all possible label sequence $\{s_r\}$	1
MMI	z	all possible label sequence $\{s_r\}$	1

This is modified from the original table in [46].

learning is applied to this generic rational-function, which includes MMI, MCE, and MPE/MWE criteria as special cases. This is significant from two perspectives. First, it provides a more precise and direct insight into the fundamental relations among MMI, MCE, and MPE/MWE criteria at the objective function level based on the common rational-function form. Second, it enables a unified GT-based parameter optimization framework that applies to MMI, MCE, MPE/MWE, and other discriminative criteria.

In [31], it was based on the objective function of the following form (rewritten using the mathematical notations of this paper for easy comparison):

$$O(\Lambda) = \frac{1}{R} \sum_{r=1}^{R} f\left(\frac{1}{\eta} \log \frac{\sum_{s_r} p^{\eta}(X_r, s_r | \Lambda) C_{\mathrm{DT}}(s_r, S_r)}{\sum_{s_r \in M_r} p^{\eta}(X_r, s_r | \Lambda)} \right) \tag{3.26}$$

where $C_{\mathrm{DT}}(s_r)$ takes the same value as in our Table 3.1. The choices of the smoothing function $f(z)$, the competing word sequences M_r, and the weight value η in (3.26) are provided in Table 3.2 for the different types of DT criteria. In Table 3.2, q is the slope of a sigmoid smoothing function.

Equation (3.26) is a generic description of the objective functions from MMI, MCE, and MPE/MWE. However, it is not at the definitive level of a unified form of a rational function. It indicates that different discriminative criteria can have a similar form of kernel and differ by the criterion dependent smoothing function $f(z)$ that modulates the kernel. The objection function of (3.26) is a sum of the smoothing functions. In the approach presented in this chapter, we show that objective functions from MMI, MCE, and MPE/MWE criteria can have a definitive common rational-function form (3.2), and for each discriminative criterion, the objective function differs only by a model-independent quantity $C_{DT}(s_1, \ldots, s_R)$.

On the other hand, as shown in Table 3.2, $f(z)$ is a nonlinear function for MPE/MWE and MCE criteria. Therefore, the original GT solution [14], while directly applicable to MMI with $f(z)$ being an identity function and z being a logarithm of rational function, is not directly applicable to the objective functions of MPE/MWE and MCE criteria. To circumvent this difficulty, the limiting procedures of [26, 27] are needed, in which the original objective function is approximated by a sequence of polynomials through Taylor series expansion (in a neighbor of the current model parameters). Based on that, the GT-based parameter optimization of [14] can be applied to each of the partial sum, a polynomial with finite number of terms. But for a nonpolynomial analytic function, the Taylor series expansion consists of infinite number of terms. It needs to justify the limiting process that the GT for polynomials with finite number of terms can be extended to the limit case as the number of terms goes to infinity, for example, the existence of a uniform bounded constant D for all partial sums of the Taylor series expansion in GT-based parameter optimization. The unified rational-function approach described in this paper departs from the work of Macherey et al. [31] and Schlüter et al. [46], because it is free from the Taylor series expansion and it maps the objective functions from MMI, MCE, and MPE/MWE criteria into a definitive rational-functional form (3.2). Therefore, the GT-based parameter optimization framework of [14] can be directly applied. Moreover, this approach allows new rational function optimization methods (e.g., the method based on reverse Jensen's inequality [22]) to be applied, upon which algorithmic properties of the parameter optimization procedure can be constructively established and justified.

· · · ·

CHAPTER 4

Discriminative Learning Algorithm for Exponential-Family Distributions

In this chapter, we describe an efficient, growth transformation (GT)-based approach to the discriminative parameter estimation problem in classifier design where each class is characterized by an exponential-family distribution discussed in Chapter 1. The next chapter extends the results here into the more difficult but practically more useful case of hidden Markov models (HMMs).

4.1 EXPONENTIAL-FAMILY MODELS FOR CLASSIFICATION

In this section, we derive the GT formulas for estimating parameters of exponential family distributions. This class of densities covers a large number of contemporary statistical models and is of important theoretical and practical interests. The derived formulas "grow" the unified rational-form discriminative training criterion $O(\Lambda)$ defined in Chapter 2. In the next chapter, we will present the derivation for the Gaussian mixture density HMM, which is widely used in modern speech recognition.

Let us start from the problem of C-class classification. Let the data of each class be i.i.d. (independent and identically distributed) that are modeled by an exponential family distribution. Although parameter estimation for this problem has a nice closed-form solution under ML training, it is more complicated under discriminative training. In the latter case, the objective function $O(\Lambda)$ is difficult to optimize directly but because it is a rational function as expressed in (3.2), we can construct the auxiliary functions of F and then V based on F (see Section 1.5.5). Optimizing $V(\Lambda; \Lambda')$ becomes a relatively easy problem and it leads to the GT formula for all types of discriminative criteria unified by (3.2).

Assume that there are R observation samples $x_r (r = 1, \ldots, R)$ in the training set and that each sample x_r is a vector with dimension D. Each sample x_r is associated with a reference label (e.g., a class index) $S_r \in \{c_i \mid i = 1, \ldots, C\}$, where C denotes the total number of classes in the task. Using the above notations, the task is considered a C-class classification problem, where each observation sample x_r is to be classified into one of the C classes.

Note that, denoting by θ_i the natural parameter vector of the exponential family distribution of the ith class, and denoting by $\Lambda = \{\theta_i\}$ the whole model parameter set, $p(x|c_i)$ takes the following form (as (1.6) in Chapter 1):

$$p(x|c_i; \Lambda) = p(x|\theta_i) = b(x) \cdot \exp\left\{\theta_i^{\mathsf{T}} T(x) - A(\theta_i)\right\} \qquad (4.1)$$

4.2   CONSTRUCTION OF AUXILIARY FUNCTIONS

According to Section 3.2.2, the objective function of discriminative training is a rational function. Following the background material presented in Section 1.5.5, we can construct the objective function of

$$O(\Lambda) = \frac{G(\Lambda)}{H(\Lambda)} \qquad (4.2)$$

in the same form of (1.26), where $G(\Lambda)$ and $H(\Lambda)$ are the numerator and denominator of (3.2). Then, the GT-based optimization algorithm constructs the auxiliary function of

$$F(\Lambda; \Lambda') = G(\Lambda) - O(\Lambda')H(\Lambda) + D \qquad (4.3)$$

where D is a quantity independent of the parameter set Λ, and Λ' denotes the parameter set obtained from the immediately previous iteration of the algorithm. The purpose of constructing (4.3) is that it is easier to optimize than (4.2). However, (4.2) is still difficult to optimize, and we desire to introduce another auxiliary function from $V(\Lambda; \Lambda')$ in (4.3). This new function is constructed by

$$V(\Lambda; \Lambda') = \int_{\chi} f(\chi, \Lambda') \log f(\chi, \Lambda) \mathrm{d}\chi \qquad (4.4)$$

where the positive, real-valued function $f(x, \Lambda) > 0$ is constructed to satisfy

$$F(\Lambda; \Lambda') = \int_{\chi} f(\chi, \Lambda) \mathrm{d}\chi \qquad (4.5)$$

Then, we have

$$\log F(\Lambda; \Lambda') - \log F(\Lambda'; \Lambda') = \log \frac{F(\Lambda; \Lambda')}{F(\Lambda'; \Lambda')}$$

$$= \log \int_{\chi} \frac{f(\chi, \Lambda')}{F(\Lambda'; \Lambda')} \frac{f(\chi, \Lambda)}{f(\chi, \Lambda')} d\chi \geq \int_{\chi} \frac{f(\chi, \Lambda')}{F(\Lambda'; \Lambda')} \log \frac{f(\chi, \Lambda)}{f(\chi, \Lambda')} d\chi$$

$$= \frac{1}{F(\Lambda'; \Lambda')} \left[\int_{\chi} f(\chi, \Lambda') \log f(\chi, \Lambda) d\chi - \int_{\chi} f(\chi, \Lambda') \log f(\chi, \Lambda') d\chi \right]$$

$$= \frac{1}{F(\Lambda'; \Lambda')} \left[V(\Lambda; \Lambda') - V(\Lambda'; \Lambda') \right] \qquad (4.6)$$

The inequality above is attributable to Jensen's inequality being applied to the concave log function. The result of (4.6) states that an increase in the auxiliary function $V(\Lambda; \Lambda')$ guarantees an increase in $\log F(\Lambda; \Lambda')$. Because logarithm is a monotonically increasing function, this also guarantees an increase of $F(\Lambda; \Lambda')$ and hence the original objective function $O(\Lambda)$. The technique that "transforms" the parameters from Λ' to Λ so as to increase or "grow" the values of the auxiliary functions and hence the value of the original objective function is called the growth-transformation (GT) technique. We now apply this GT technique to the exponential-family distribution with the unified discriminative optimization criterion formulated in (3.2).

4.3   GT LEARNING FOR EXPONENTIAL-FAMILY DISTRIBUTIONS

In this section, we derive the GT formula for general exponential-family distributions. The formula "grows" the unified discriminative training criterion $O(\Lambda)$. As discussed above, $O(\Lambda)$ is difficult to optimize directly but because it is a rational function as expressed in (3.2), we can construct the auxiliary functions of (1) F and then (2) V based on F. Optimizing $V(\Lambda; \Lambda')$ becomes a relatively easier problem and it leads to the final GT formula for all types of discriminative criteria unified by (3.2). In the next section, we will present the derivation for two specific exponential-family distributions — multinomial distribution and Gaussian distribution.

In the rational function of

$$O(\Lambda) = \frac{G(\Lambda)}{H(\Lambda)} \qquad (4.7)$$

as the unified form of the discriminative objective function (3.2) for maximum mutual information, minimum classification error, and minimum phone error/minimum word error, we have

$$G(\Lambda) = \sum_s p(X, s|\Lambda)C(s) \text{ and } H(\Lambda) = \sum_s p(X, s|\Lambda) \qquad (4.8)$$

where we use $s = s_1, \ldots, s_R$ to denote the class label (including correct or incorrect labels) for each of the R training tokens, respectively, and use $X = x_1, \ldots, x_R$ to denote the observation samples for these R training tokens. Note that each observation sample is a feature vector.

For the auxiliary function of

$$F(\Lambda; \Lambda') = G(\Lambda) - O(\Lambda')H(\Lambda) + D \qquad (4.9)$$

we substitute (4.8) into (4.9) to obtain the new auxiliary function

$$F(\Lambda; \Lambda') = \sum_s p(X, s|\Lambda)C(s) - O(\Lambda') \sum_s p(X, s|\Lambda) + D$$

$$= \sum_s p(X, s|\Lambda) \left[C(s) - O(\Lambda') \right] + D \qquad (4.10)$$

The main terms in the auxiliary function $F(\Lambda; \Lambda')$ above can be interpreted as the average deviation of the accuracy count.

Because $p(s)$ is the prior probability of s, it is irrelevant for optimizing Λ. Using $p(X, s|\Lambda) = p(s) \cdot p(X|s, \Lambda)$, we obtain

$$F(\Lambda; \Lambda') = \sum_s [C(s) - O(\Lambda')]p(s)p(X|s, \Lambda) + D$$

$$= \sum_s \sum_\chi \left[\Gamma(\Lambda') + d(s) \right] p(\chi|s, \Lambda) \qquad (4.11)$$

where

$$\Gamma(\Lambda') = \delta(\chi, X)p(s) \left[C(s) - O(\Lambda') \right] \qquad (4.12)$$

In (4.10), $D = \sum_s d(s)$ is a quantity independent of the parameter set Λ. In (4.12), $\delta(\chi, X)$ is the Kronecker delta function for discrete valued observations, and χ represents the entire data space where X belongs. The summation over this data space is introduced here for accommodating the parameter-independent constant D; that is, $\sum_s \sum_\chi d(s)p(\chi|s, \Lambda) = \sum_s d(s) = D$ is a Λ-independent constant.

In the case where the observation vector is continuous valued, the summation operation above will be replaced with integration, and $\delta(\chi, X)$ in (4.12) needs to be a Dirac delta function.

We now proceed to construct the new auxiliary function of (4.4). To achieve this, we first identify from (4.5) and (4.11) that

$$f(\chi, s, \Lambda) = \left[\Gamma(\Lambda') + d(s) \right] p(\chi|s, \Lambda)$$

To ensure that $f(\chi, s, \Lambda)$ above is positive, $d(s)$ should be selected to be sufficiently large so that $\Gamma(\Lambda') + d(s) > 0$ (note that $p(\chi|s, \Lambda)$ is nonnegative). This issue will be discussed in greater details in Section 5.4.

Then, using (4.4), we have

$$
\begin{aligned}
V(\Lambda; \Lambda') &= \sum_{s}\sum_{\chi} \left[\Gamma(\Lambda') + d(s)\right] p(\chi | s, \Lambda') \log \left\{ \underbrace{\left[\Gamma(\Lambda') + d(s)\right]}_{\text{optimization - indept}} p(\chi | s, \Lambda) \right\} \\
&= \sum_{s}\sum_{\chi} \left[\Gamma(\Lambda') + d(s)\right] p(\chi | s, \Lambda') \log p(\chi | s, \Lambda') + \text{Const.} \\
&= \sum_{s} p(X, s|\Lambda') \left(C(s) - O(\Lambda')\right) \log p(X|s, \Lambda') \\
&\quad + \sum_{s}\sum_{\chi} d(s) p(\chi|s, \Lambda') \log p(\chi | s, \Lambda') + \text{Const.}
\end{aligned}
\tag{4.13}
$$

The key reason which makes this new auxiliary function (4.13) easier to optimize than that in (4.11) is the new logarithm in $\log p(X| s, \Lambda)$, which was absent in (4.11). As for the case of ML learning case for exponential-family distributions, this enables drastic simplification of the new auxiliary function of $V(\Lambda; \Lambda')$, which we outline below.

We first ignore the optimization-independent constant in (4.13), and divide $V(\Lambda; \Lambda')$ by another optimization-independent quantity, $p(X|\Lambda')$, in order to convert the joint probability $p(X, s |\Lambda')$ to the posterior probability $p(s, X|\Lambda')$. We then obtain an equivalent auxiliary function of

$$
U(\Lambda; \Lambda') = \overbrace{\sum_{s} p(s|X, \Lambda') \left(C(s) - O(\Lambda')\right) \log p(X|s, \Lambda')}^{\text{term - I}} \\
+ \underbrace{\sum_{s}\sum_{\chi} d'(s) p(\chi | s, \Lambda') \log p(\chi | s, \Lambda)}_{\text{term - II}}
\tag{4.14}
$$

where

$$
d'(s) = d(s)/p(X|\Lambda')
\tag{4.15}
$$

Note that $X = X_1, \ldots, X_R$ is a large aggregate of all training data with R independent tokens. For each token X_r, it is independent of each other and it depends only on the rth label. Hence, $\log p(X| s, \Lambda)$ can be decomposed, enabling simplification of both term-I and term-II in (4.14). We now elaborate on the simplification of these two terms.

$$\text{term - I} = \sum_s p(s|X,\Lambda')\left(C(s) - O(\Lambda')\right) \sum_{r=1}^{R} \log p(x_r|s_r,\Lambda)$$

$$= \sum_s p(s|X,\Lambda')\left(C(s) - O(\Lambda')\right) \sum_{\substack{r=1 \\ s_r=c_i}}^{R} \sum_{i=1}^{C} \log p(x_r|c_i,\Lambda) \qquad (4.16)$$

The simplification process for the second term in (4.14) is below (using the notations $\tilde{\mathbf{s}} = s_1,$ $\dots, s_{r-1}, s_{r+1}, \dots, s_R, \tilde{\chi} = \chi_1, \dots, \chi_{r-1}, \chi_{r+1}, \dots, \chi_R$):

$$\text{term - II} = \sum_s d'(s) \sum_{\chi_1,\dots,\chi_R} p\left(\chi_1,\dots,\chi_R|s,\Lambda'\right) \sum_{r=1}^{R} \log p\left(\chi_r|s_r,\Lambda\right)$$

$$= \sum_s d'(s) \sum_{r=1}^{R} \sum_{\chi_r} p(\chi_r|s_r,\Lambda') \underbrace{\sum_{\tilde{\chi}} p\left(\tilde{\chi}|\tilde{s},\Lambda'\right)}_{=1} \log p\left(\chi_r|s_r,\Lambda\right)$$

$$= \sum_{r=1}^{R} \sum_{i=1}^{I} d(r,i) \sum_{\chi_{r,t}} p(\chi_r|c_i,\Lambda') \log p\left(\chi_r|c_i,\Lambda\right) \qquad (4.17)$$

where

$$d(r,i) = \sum_{s,\, s_r=c_i} d'(s) \qquad (4.18)$$

Substituting (4.16) and (4.17) into (4.14), and using (4.1), we have:

$$U(\Lambda;\Lambda') = \sum_s p\left(s|X,\Lambda'\right)\left(C(s) - O(\Lambda')\right) \sum_{\substack{r=1 \\ s_r=c_i}}^{R} \sum_{i=1}^{C} \log p(x_r|\theta_i)$$

$$+ \sum_{r=1}^{R} \sum_{i=1}^{C} d(r,i) \sum_{\chi_{r,t}} p\left(\chi_r|c_i,\Lambda'\right) \log p\left(\chi_r|\theta_i\right) \qquad (4.19)$$

Because $p(\cdot|\theta_i)$ is an exponential density and therefore its logarithm is a linear function of the data, $U(\Lambda;\Lambda')$ becomes ready to be maximized, which we proceed below.

Setting, $\dfrac{\partial U(\Lambda;\Lambda')}{\partial \theta_i} = 0, i = 1, \dots, C$, we obtain

$$0 = \sum_s p\left(s|X,\Lambda'\right)\left(C(s) - O(\Lambda')\right) \sum_{\substack{r=1 \\ s_r = c_i}}^{R}\left(T(x_r) - \frac{\partial A(\theta_i)}{\partial \theta_i}\right)$$

$$+ \sum_{r=1}^{R} \mathrm{d}(r,i)\sum_{\chi_{r,t}} p\left(\chi_r|c_i,\Lambda'\right)\left(T(\chi_r) - \frac{\partial A(\theta_i)}{\partial \theta_i}\right)$$

with the constraints:

$$\sum_{\chi_{r,t}} p\left(\chi_r|c_i,\Lambda'\right) = 1$$

$$\sum_{\chi_{r,t}} p\left(\chi_r|c_i,\Lambda'\right) T(\chi_r) = \mathbb{E}_{p(\chi|\theta_i')}[T(\chi)]$$

Using

$$D_i = \sum_{r=1}^{R} \mathrm{d}(r,i)$$

we obtain the solution that satisfies

$$\frac{\partial A(\theta_i)}{\partial \theta_i} = \frac{\displaystyle\sum_s p\left(s|X,\Lambda'\right)\left(C(s) - O(\Lambda')\right)\sum_{\substack{r=1 \\ s_r = c_i}}^{R} T(x_r) + D_i \cdot \mathbb{E}_{p(\chi|\theta_i')}[T(\chi)]}{\displaystyle\sum_s p(s|X,\Lambda')\left(C(s) - O(\Lambda')\right)\sum_{\substack{r=1 \\ s_r = c_i}}^{R} 1 + D_i} \tag{4.20}$$

If we define

$$\Delta\gamma(i,r) = \sum_s p\left(s|X,\Lambda'\right)\left(C(s) - O(\Lambda')\right)\delta(s_r = c_i) \tag{4.21}$$

Then we can rewrite (4.20) as

$$\frac{\partial A(\theta_i)}{\partial \theta_i} = \frac{\displaystyle\sum_{r=1}^{R} \Delta\gamma(i,r)T(x_r) + D_i \cdot \mathbb{E}_{p(\chi|\theta_i')}[T(\chi)]}{\displaystyle\sum_{r=1}^{R} \Delta\gamma(i,r) + D_i} \tag{4.22}$$

Based on (4.22), we will present details in deriving the model estimation formulas for two important exponential-family distributions in the next section. These are multinomial distribution and Gaussian distributions. The former is commonly used to model discrete distribution and the latter is widely used to model continuous random variables. Equation (4.22) is also applicable to all other members of the exponential family.

4.4   ESTIMATION FORMULAS FOR TWO EXPONENTIAL-FAMILY DISTRIBUTIONS

4.4.1   Multinomial Distribution

In this section we discuss the discriminative training formulas when $p(x|\theta_i)$ is a single-observation multinomial distribution. Readers are referred to Section 1.4.3 for a general introduction of multinomial distribution and its properties.

For the single observation multinomial distribution of the ith class, its standard form is

$$p(x|v_i) = \prod_{k=1}^{K} v_{i,k}^{x(k)}$$

where $x = [x(1), ..., x(K)]^{\mathrm{T}}$ is a K-dimensional observation vector and $v_i = [v_{i,1}, ..., v_{i,K}]^{\mathrm{T}}$ is the K-dimensional parameter vector.

The exponential-family form of the is single observation multinomial distribution

$$p(x|\theta_i) = h(x) \cdot \exp\left\{ \theta_i^{\mathrm{T}} T(x) - A(\theta_i) \right\}$$

where

$$T(x) = \tilde{x} = [x(1),...,x(K-1)]^{\mathrm{T}} \tag{4.23}$$

with $\tilde{x}$ being the observation vector that contains the first $K-1$ elements of x. Given the above definition, according to properties of the multinomial distribution, we have the following

$$\mathbb{E}_{p(\chi|\theta_i')}[T(\chi)] = \mathbb{E}_{p(\chi|v_i')}[\tilde{\chi}] = \tilde{v}_i' \tag{4.24}$$

where $\tilde{v}_i' = [v_{i,1}', ..., v_{i,K-1}']^{\mathrm{T}}$ is an parameter vector that contains only the first $K-1$ parameters.

Substituting (1.16), (4.23), and (4.24) into (4.22), and denote by $\tilde{v}_i' = [v_{i,1}', ..., v_{i,K-1}']^{\mathrm{T}}$, we have

$$\tilde{v}_i = \frac{\sum_{r=1}^{R} \Delta\gamma(i,r)\tilde{x}_r + D_i \cdot \tilde{v}_i'}{\sum_{r=1}^{R} \Delta\gamma(i,r) + D_i} \tag{4.25}$$

By summing both sides of (4.25) over $k = 1, \ldots, K - 1$, we have

$$\sum_{k=1}^{K-1} v_{i,k} = \frac{\sum_{r=1}^{R} \Delta\gamma(i,r) \sum_{k=1}^{K-1} x_{r,k} + D_i \cdot \sum_{k=1}^{K-1} v'_{i,k}}{\sum_{r=1}^{R} \Delta\gamma(i,r) + D_i}$$

from which, we obtain

$$v_{i,K} = 1 - \sum_{k=1}^{K-1} v_{i,k}$$

$$= 1 - \frac{\sum_{r=1}^{R} \Delta\gamma(i,r) \sum_{k=1}^{K-1} x_{r,k} + D_i \cdot \sum_{k=1}^{K-1} v'_{i,k}}{\sum_{r=1}^{R} \Delta\gamma(i,r) + D_i}$$

$$= \frac{\sum_{r=1}^{R} \Delta\gamma(i,r) \left(1 - \sum_{k=1}^{K-1} x_{r,k}\right) + D_i \cdot \left(1 - \sum_{k=1}^{K-1} v'_{i,k}\right)}{\sum_{r=1}^{R} \Delta\gamma(i,r) + D_i}$$

$$= \frac{\sum_{r=1}^{R} \Delta\gamma(i,r) x_{r,K} + D_i \cdot v'_{i,K}}{\sum_{r=1}^{R} \Delta\gamma(i,r) + D_i} \tag{4.26}$$

Combining (4.25) and (4.26), we have the GT estimation formula for multinomial distribution:

$$v_i = \frac{\sum_{r=1}^{R} \Delta\gamma(i,r) x_r + D_i \cdot v'_i}{\sum_{r=1}^{R} \Delta\gamma(i,r) + D_i} \tag{4.27}$$

4.4.2   Multivariate Gaussian Distribution

In this section, we will derive and present discriminative training formulas when $p(\mathbf{x}|\theta_i)$ is a multivariate Gaussian distribution. Readers are referred to Section 1.4.3 for a general introduction of multivariate Gaussian distribution and its properties.

For a multivariate Gaussian distribution of the ith class, its standard form is

$$p(x|\lambda_i) = N(x|\mu_i, \Sigma_i) = \frac{1}{(2\pi)^{\frac{D}{2}}|\Sigma_i|^{\frac{1}{2}}} \exp\left\{-\frac{1}{2}(x-\mu_i)^T \Sigma_i^{-1}(x-\mu_i)\right\} \qquad (4.28)$$

and its exponential family form is

$$p(x|\theta_i) = h(x) \cdot \exp\left\{\theta_i^T T(x) - A(\theta_i)\right\}$$

where

$$T(x) = \begin{bmatrix} x_1 \\ x_2 \end{bmatrix} = \begin{bmatrix} x \\ \text{Vec}\left(xx^T\right) \end{bmatrix} \qquad (4.29)$$

Given the above definition, according to the property of the multivariate Gaussian distribution, we have

$$\mathbb{E}_{p(\chi|\theta_i')}\left[T(\chi)\right] = \mathbb{E}_{p(\chi|\mu_i', \Sigma_i')} \begin{bmatrix} x \\ \text{Vec}\left(xx^T\right) \end{bmatrix} = \begin{bmatrix} \mu_i' \\ \text{Vec}\left(\Sigma_i' + \mu_i'\mu_i'^T\right) \end{bmatrix} \qquad (4.30)$$

Substituting (1.22), (1.23), (4.29), and (4.30) into (4.22), we finally obtain

$$\begin{bmatrix} \mu_i \\ \text{Vec}\left(\mu_i\mu_i^T + \Sigma_i\right) \end{bmatrix} = \begin{bmatrix} \dfrac{\displaystyle\sum_{r=1}^{R}\Delta\gamma(i,r)x_r + D_i \cdot \mu_i'}{\displaystyle\sum_{r=1}^{R}\Delta\gamma(i,r) + D_i} \\[4ex] \dfrac{\displaystyle\sum_{r=1}^{R}\Delta\gamma(i,r)\text{Vec}\left(x_r x_r^T\right) + D_i \cdot \text{Vec}\left(\Sigma_i' + \mu_i'\mu_i'^T\right)}{\displaystyle\sum_{r=1}^{R}\Delta\gamma(i,r) + D_i} \end{bmatrix}$$

After rearrangement and canceling out the Vec() function at both sides, we can obtain the parameter updating formulas as

$$\mu_i = \frac{\displaystyle\sum_{r=1}^{R}\Delta\gamma(i,r)x_r + D_i \cdot \mu_i'}{\displaystyle\sum_{r=1}^{R}\Delta\gamma(i,r) + D_i} \qquad (4.31)$$

$$\Sigma_i = \frac{\sum\limits_{r=1}^{R} \Delta\gamma(i,r)x_r x_r^{\mathrm{T}} + D_i \cdot \left[\Sigma_i' + \mu_i' \mu_i'^{\mathrm{T}}\right]}{\sum\limits_{r=1}^{R} \Delta\gamma(i,r) + D_i} - \mu_i \mu_i^{\mathrm{T}} \tag{4.32}$$

Equations (4.27), (4.31), and (4.32) give the discriminative training formula for the multinomial distribution and Gaussian distribution. The computation of $\Delta\gamma(i, r)$ will be presented in greater details in Chapter 6, and the issues of setting constant D_i is discussed in Section 5.4.

·   ·   ·   ·

CHAPTER 5

Discriminative Learning Algorithm for Hidden Markov Model

In this chapter, we extend the growth transformation (GT)-based approach to the discriminative parameter estimation problem from the stationary probability model characterized by the exponential-family distributions discussed in Chapter 4 to the nonstationary probability model. The nonstationarity discussed here is characterized by the Markov chain that underlies the hidden Markov model (HMM), where the emission probabilities in the HMM can be represented by any member in the exponential-family distributions as well as by the mixture Gaussian distribution. Specifically, in the algorithm derivation discussed in this chapter, we only use the example for the Gaussian HMM. It can be easily generalized to other types of continuous-density HMMs (CDHMMs) by using some of the derivation steps discussed in Chapter 4. To make the coverage of HMM complete, we first discuss the discriminative parameter estimation problem in classifier design where each class is characterized by a discrete HMM. This is then followed by a discussion of CDHM.

5.1 ESTIMATION FORMULAS FOR DISCRETE HMM

In this section, we derive the GT estimation formulas for the discrete HMM's parameters — $\Lambda = \{\{a_{ij}\}, \{b_i(k)\}\}$ for transition probabilities and emitting probabilities. The formula "grows" the unified discriminative training criterion $O(\Lambda)$. In the next section, we will present the derivation for the CDHMM. In both cases, $O(\Lambda)$ is difficult to optimize directly but because it is a rational function as expressed in (3.2), we can construct the auxiliary functions of (1) F and then (2) V based on F. Optimizing $V(\Lambda; \Lambda')$ becomes a relatively easier problem and it leads to the final GT formulas for all types of discriminative criteria unified by (3.2).

For the discrete HMM, $X = X_1, \ldots, X_R$ is used to denote observation sequences with discrete indexes. That is, at time t for token r, the observation $x_{r,t}$ is an index belonging to the set $[1, 2, \ldots, K]$, where K is the size of the index set.

5.1.1   Constructing Auxiliary Function $F(\Lambda; \Lambda')$

Starting from (4.11), we have the generic auxiliary function of

$$F(\Lambda;\Lambda') = \sum_s p(X,s|\Lambda)\left[C(s) - O(\Lambda')\right] + D$$

which, for an HMM, becomes

$$F(\Lambda;\Lambda') = \sum_s \sum_q p(X,q,s|\Lambda)\left[C(s) - O(\Lambda')\right] + D \qquad (5.1)$$

where q is the HMM state sequence, and $s = s_1, \ldots, s_R$ is the label (e.g., word or phone) sequence for all R training tokens (including correct or incorrect sentences). The same interpretation as in Chapter 4 can be given to the main terms in the auxiliary function $F(\Lambda;\Lambda')$ above as the average deviation of the accuracy count.

Because $p(s)$ depends on the language model, it is irrelevant for optimizing Λ. We therefore can have the following decomposition: $p(X,q,s|\Lambda) = p(s) \cdot p(X, q|s, \Lambda)$. Hence,

$$F(\Lambda;\Lambda') = \sum_s \sum_q [C(s) - O(\Lambda')]p(s)p(X,q|s,\Lambda) + D$$

$$= \sum_s \sum_q \sum_\chi [\Gamma(\Lambda') + d(s)] p(\chi, q|s, \Lambda) \qquad (5.2)$$

where

$$\Gamma(\Lambda') = \delta(\chi,X)p(s)\left[C(s) - O(\Lambda')\right] \qquad (5.3)$$

As before, $D = \sum_s d(s)$ is a quantity independent of the parameter set Λ. In (5.3), $\delta(\chi, X)$ is the Kronecker delta function, in which χ represents the entire data space where X is in. The summation over this data space is introduced here again for accommodating the parameter-independent constant D, that is, $\sum_s \sum_g \sum_\chi d(s)p(\chi, q \mid s, \Lambda) = \sum_s d(s) = D$ is a Λ-independent constant (an idea originally proposed in [17] for HMM).

5.1.2.   Constructing Auxiliary Function $V(\Lambda; \Lambda')$

We now desire to construct the new auxiliary function $V(\Lambda; \Lambda')$ using (5.2), in the same way as we did in Chapter 4. We first identify

$$f(\chi,q,s,\Lambda) = [\Gamma(\Lambda') + d(s)] p(\chi,q|s,\Lambda)$$

as before. Again, to ensure that $f(\chi, q, s, \Lambda)$ above is positive, $d(s)$ should be selected to be sufficiently large so that $\Gamma(\Lambda') + d(s) > 0$ (note $p(\chi, q \mid s, \Lambda)$ in (5.2) is nonnegative). Then, using (4.4) again, we have

$$V(\Lambda;\Lambda') = \sum_{q}\sum_{s}\sum_{\chi} \left[\Gamma(\Lambda') + d(s)\right] p\left(\chi, q | s, \Lambda'\right) \log \left\{ \underbrace{\left[\Gamma(\Lambda') + d(s)\right] p\left(\chi, q | s, \Lambda\right)}_{\text{optimization - indept}} \right\}$$

$$= \sum_{q}\sum_{s}\sum_{\chi} \left[\Gamma(\Lambda') + d(s)\right] p(\chi, q | s, \Lambda') \log \left(\chi, q | s, \Lambda\right) + \text{Const.}$$

$$= \sum_{q}\sum_{s} p(X, q, s | \Lambda') \left(C(s) - O(\Lambda')\right) \log p\left(X, q | s, \Lambda\right)$$

$$+ \sum_{q}\sum_{s}\sum_{\chi} d(s) p\left(\chi, q | s, \Lambda\right) \log p\left(\chi, q | s, \Lambda\right) + \text{Const.} \tag{5.4}$$

5.1.3   Simplifying Auxiliary Function $V(\Lambda; \Lambda')$

After ignoring optimization-independent constant in (5.4), we divide $V(\Lambda; \Lambda')$ by $p(X|\Lambda')$ to convert the joint probability $p(X, q, s | \Lambda')$ to the posterior probability $p(q, s | X, \Lambda') = p(s | X, \Lambda')$ $p(q | X, s, \Lambda')$. The equivalent auxiliary function then becomes

$$U(\Lambda;\Lambda') = \sum_{q}\sum_{s} p\left(s | X, \Lambda'\right) p\left(q | X, s, \Lambda'\right) \left(C(s) - O(\Lambda')\right) \log p\left(X, q | s, \Lambda\right)$$

$$+ \sum_{q}\sum_{s}\sum_{\chi} d'(s) p\left(\chi, q | s, \Lambda'\right) \log p\left(\chi, q | s, \Lambda\right) \tag{5.5}$$

where

$$d'(s) = d(s) / p\left(X | \Lambda'\right) \tag{5.6}$$

Because X depends only on the HMM state sequence q, we have $p(X, q | s, \Lambda) = p(q | s, \Lambda) \cdot p(X | q, \Lambda)$. Therefore, $U(\Lambda; \Lambda')$ can be further decomposed to four terms below:

$$U(\Lambda;\Lambda') = \overbrace{\sum_{q}\sum_{s} p\left(s | X, \Lambda'\right) p\left(q | X, s, \Lambda'\right) \left(C(s) - O(\Lambda')\right) \log p\left(X | q, \Lambda\right)}^{\text{term - I}}$$

$$+ \underbrace{\sum_{q}\sum_{s}\sum_{\chi} d'(s) p\left(\chi, q | s, \Lambda'\right) \log p\left(\chi | q, \Lambda\right)}_{\text{term - II}}$$

$$+ \overbrace{\sum_{q}\sum_{s} p\left(s | X, \Lambda'\right) p\left(q | X, s, \Lambda'\right) \left(C(s) - O(\Lambda')\right) \log p\left(q | s, \Lambda\right)}^{\text{term - III}}$$

$$+ \underbrace{\sum_{q}\sum_{s}\sum_{\chi} d'(s) p\left(\chi, q | s, \Lambda'\right) \log p\left(q | s, \Lambda\right)}_{\text{term - IV}} \tag{5.7}$$

Note the two new terms above compared with the corresponding auxiliary function (4.14) in Chapter 4. Here, $X = X_1, \ldots, X_R$ is a large aggregate of all training data with R independent sentence tokens. For each token $X_r = x_{r,1}, \ldots, x_{r,T_r}$, the observation vector $x_{r,t}$ is independent of each other and it depends only on the HMM state at time t. Hence, $\log p(X|q, \Lambda)$ can be decomposed, enabling simplification of both term-I and term-II in (5.7). To simplify term-III and term-IV in (5.7), we decompose $\log p(q|s, \Lambda)$ based on the property of the first-order HMM — the state at time t depends only on state at time $t-1$. We now elaborate on the simplification of all four terms in (5.7).

For term-I, we first define

$$\gamma_{i,r,s_r}(t) \triangleq \sum_{q, q_{r,t}=i} p\left(q|X, s, \Lambda'\right) = p\left(q_{r,t} = i|X, s, \Lambda'\right) = p\left(q_{r,t} = i|X_r, s_r, \Lambda'\right) \qquad (5.8)$$

The last equality comes from the fact that the sentence tokens in the training set are independent of each other. $\gamma_{i,r,s_r}(t)$ is the occupation probability of state i at time t, given the label sequence s_r and observation sequence X_r, and an efficient forward–backward algorithm exists to compute it [43]. Using the definition of (5.8) and assuming that the HMM state index is from 1 to I, we have

$$\text{term - I} = \sum_s p\left(s|X, \Lambda'\right)\left(C(s) - O(\Lambda')\right) \sum_q p\left(q|X, s, \Lambda'\right) \sum_{r=1}^{R} \sum_{t=1}^{T_r} \log p\left(x_{r,t}|q_{r,t}, \Lambda\right)$$

$$= \sum_s p\left(s|X, \Lambda'\right)\left(C(s) - O(\Lambda')\right) \sum_{r=1}^{R} \sum_{t=1}^{T_r} \sum_{i=1}^{I} \sum_{q, q_{r,t}=i} p\left(q|X, s, \Lambda'\right) \log p\left(x_{r,t}|q_{r,t} = i, \Lambda\right)$$

$$= \sum_s p\left(s|X, \Lambda'\right)\left(C(s) - O(\Lambda')\right) \sum_{r=1}^{R} \sum_{t=1}^{T_r} \sum_{i=1}^{I} \gamma_{i,r,s_r}(t) \log p\left(x_{r,t}|q_{r,t} = i, \Lambda\right) \qquad (5.9)$$

The simplification process for the second term in (5.7) is as follows. Using the notations $\tilde{q} = q_{1,1}, \ldots, q_{r,t-1}, q_{r,t+1}, \ldots, q_{R,T_R}, \tilde{X} = X_{1,1}, \ldots, X_{r,t-1}, X_{r,t+1}, \ldots, X_{R,T_R}$, we have

$$\text{term - II} = \sum_s d'(s) \sum_{q_{1,1},\dots,q_{R,T_R}} \sum_{\chi_{1,1},\dots,\chi_{R,T_R}} p\left(\chi_{1,1},\dots,\chi_{R,T_R},q_{1,1},\dots,q_{R,T_R}\,\Big|\,s,\Lambda'\right)\sum_{r=1}^{R}\sum_{t=1}^{T_r}\log p\left(\chi_{r,t}\Big|q_{r,t},\Lambda\right)$$

$$= \sum_s d'(s) \sum_{r=1}^{R}\sum_{t=1}^{T_r}\sum_{q_{r,t}}\sum_{\chi_{r,t}} p\left(\chi_{r,t},q_{r,t}\,\Big|\,s,\Lambda'\right)\underbrace{\sum_{\tilde{q}}\sum_{\tilde{\chi}} p(\tilde{\chi},\tilde{q}|\chi_{r,t},q_{r,t},s,\Lambda')}_{=1}\log p\left(\chi_{r,t}\Big|q_{r,t},\Lambda\right)$$

$$= \sum_s d'(s) \sum_{r=1}^{R}\sum_{t=1}^{T_r}\sum_{q_{r,t}}\sum_{\chi_{r,t}} p\left(\chi_{r,t},q_{r,t}\,\Big|\,s,\Lambda'\right)\log p\left(\chi_{r,t}\Big|q_{r,t},\Lambda\right)$$

$$= \sum_{r=1}^{R}\sum_{t=1}^{T_r}\sum_{i=1}^{I}\sum_{\chi_{r,t}}\sum_{s} d'(s)p(q_{r,t}=i|s,\Lambda')p(\chi_{r,t}|q_{r,t}=i,\Lambda')\log\ p(\chi_{r,t}|q_{r,t}=i,\Lambda)$$

$$= \sum_{r=1}^{R}\sum_{t=1}^{T_r}\sum_{i=1}^{I} d(r,t,i)\sum_{\chi_{r,t}} p\left(\chi_{r,t}\Big|q_{r,t}=i,\Lambda'\right)\log p\left(\chi_{r,t}\Big|q_{r,t}=i,\Lambda\right) \tag{5.10}$$

where

$$d(r,t,i) = \sum_s d'(s)p\left(q_{r,t}=i\,\Big|\,s,\Lambda'\right) \tag{5.11}$$

To simplify term-III in (5.7), we first define

$$\xi_{i,j,r,s_r}(t) \triangleq \sum_{q:q_{r,t-1}=i,q_{r,t}=j} p\left(q|X,s,\Lambda'\right) = p\left(q_{r,t-1}=i,q_{r,t}=j|X,s,\Lambda'\right)$$

$$= p\left(q_{r,t-1}=i,q_{r,t}=j|X_r,s_r,\Lambda'\right) \tag{5.12}$$

which is the posterior probability of staying at state i at time $t-1$ and staying at state j at time t, given the label sequence s_r and observation sequence X_r. An efficient forward–backward algorithm exists to compute this posterior probability in a standard way [43]. Then, we decompose $p(q\,|\,s,\Lambda)$ as follows:

$$p(q|s,\Lambda) = \prod_{r=1}^{R} p\left(q_{r,1},\dots,q_{r,T_r}|s_r,\Lambda\right) = \prod_{r=1}^{R}\prod_{t=1}^{T_r} a_{q_{r,t-1},q_{r,t}}$$

This leads to the following simplifications:

$$\text{term - III} = \sum_s p\left(s|X,\Lambda'\right)\left(C(s) - O(\Lambda')\right)\sum_q p\left(q|X,s,\Lambda'\right)\sum_{r=1}^{R}\sum_{t=1}^{T_r}\log a_{q_{r,t-1},q_{r,t}}$$

$$= \sum_s p\left(s|X,\Lambda'\right)\left(C(s) - O(\Lambda')\right)\sum_{r=1}^{R}\sum_{t=1}^{T_r}\sum_{i=1}^{I}\sum_{j=1}^{I}\sum_{q,q_{r,t-1}=i,q_{r,t}=j} p\left(q|X,s,\Lambda'\right)\log a_{i,j}$$

$$= \sum_s p\left(s|X,\Lambda'\right)\left(C(s) - O(\Lambda')\right)\sum_{r=1}^{R}\sum_{t=1}^{T_r}\sum_{i=1}^{I}\sum_{j=1}^{I}\xi_{i,j,r,s_r}(t)\log a_{i,j} \qquad (5.13)$$

and

$$\text{term - IV} = \sum_s d'(s)\sum_q\sum_\chi p\left(\chi,q|s,\Lambda'\right)\sum_{r=1}^{R}\sum_{t=1}^{T_r}\log a_{q_{r,t-1},q_{r,t}}$$

$$= \sum_s d'(s)\sum_{r=1}^{R}\sum_{t=1}^{T_r}\sum_q\sum_\chi p\left(\chi,q|s,\Lambda'\right)\log a_{q_{r,t-1},q_{r,t}}$$

$$= \sum_s d'(s)\sum_{r=1}^{R}\sum_{t=1}^{T_r}\sum_{q_{r,t-1}}\sum_{q_{r,t}} p\left(q_{r,t-1},q_{r,t}|s,\Lambda'\right)\log a_{q_{r,t-1},q_{r,t}}$$

$$= \sum_s d'(s)\sum_{r=1}^{R}\sum_{t=1}^{T_r}\sum_{i=1}^{I}\sum_{j=1}^{I} p\left(q_{r,t-1}=i|s,\Lambda'\right)p\left(q_{r,t}=j|q_{r,t-1}=i,s,\Lambda'\right)\log a_{i,j}$$

$$= \sum_{r=1}^{R}\sum_{t=1}^{T_r}\sum_{i=1}^{I} d(r,t-1,i)\sum_{j=1}^{I} a'_{i,j}\log a_{i,j} \qquad (5.14)$$

where $a'_{i,j} = p(q_{r,t}=j|q_{r,t-1}=i,s,\Lambda')$ is the transition probability from the previous GT iteration.

Substituting (5.19), (5.10), (5.13), and (5.14) into (5.7), and denoting the emitting probability by $b_i(x_{r,t}) = p(x_{r,t}|q_{r,t}=i,\Lambda)$ and $b'_i(x_{r,t}) = p(x_{r,t}|q_{r,t}=i,\Lambda')$, we obtain the decomposed and simplified objective function:

$$U(\Lambda;\Lambda') = U_1(\Lambda;\Lambda') + U_2(\Lambda;\Lambda') \qquad (5.15)$$

where

$$U_1(\Lambda;\Lambda') = \sum_{r=1}^{R}\sum_{t=1}^{T_r}\sum_{i=1}^{I}\sum_s \left(s|X,\Lambda'\right)\left(C(s) - O(\Lambda')\right)\gamma_{i,r,s_r}(t)\log b_i(x_{r,t})$$

$$+ \sum_{r=1}^{R}\sum_{t=1}^{T_r}\sum_{i=1}^{I} d(r,t,i)\sum_{\chi_{r,t}} b'_i(\chi_{r,t})\log b_i(\chi_{r,t}) \qquad (5.16)$$

$$U_2(\Lambda;\Lambda') = \sum_{r=1}^{R}\sum_{t=1}^{T_r}\sum_{i=1}^{I}\sum_{j=1}^{I}\sum_{s} p\left(s\,|X,\Lambda'\right)\left(C(s) - O(\Lambda')\right)\xi_{i,j,r,s_r}(t)\log a_{i,j}$$

$$+ \sum_{r=1}^{R}\sum_{t=1}^{T_r}\sum_{i=1}^{I} d(r,t-1,i)\sum_{j=1}^{I} a'_{i,j}\log a_{i,j} \tag{5.17}$$

In (5.15), $U_1(\Lambda; \Lambda')$ is relevant to optimizing the emitting probability $b_i(k)$, and $U_2(\Lambda; \Lambda')$ is relevant to optimizing the transition probability $a_{i,j}$.

5.1.4   GT by Optimizing Auxiliary Function $U(\Lambda; \Lambda')$

To optimize the discrete distribution $b_i(k) = p(x_{r,t} = k\,|q_{r,t} = i, \Lambda)$, $k = 1, 2, \ldots, K$, where the constraint $\sum_{k=1}^{K} b_i(k) = 1$ is imposed, we apply the Lagrange multiplier method by constructing

$$W_1(\Lambda;\Lambda') = U_1(\Lambda;\Lambda') + \sum_{i=1}^{I} \lambda_i \left(\sum_{k=1}^{K} b_i(k) - 1 \right)$$

Setting $\dfrac{\partial W_1(\Lambda; \Lambda')}{\partial \lambda_i} = 0$ and $\dfrac{\partial W_1(\Lambda; \Lambda')}{\partial b_i(k)} = 0$, $k = 1, \ldots, K$, we have the following $K + 1$ equations:

$$\sum_{k=1}^{K} b_i(k) - 1 = 0$$

$$0 = \lambda_i b_i(k) + \sum_{r=1}^{R}\;\sum_{\substack{t=1 \\ s.t.x_{r,t}=k}}^{T_r}\overbrace{\sum_{s} p\left(s\,|X,\Lambda'\right)\left(C(s) - O(\Lambda')\right)\gamma_{i,r,s_r}(t)}^{\Delta\gamma(i,r,t)}\cdot$$

$$+ \sum_{r=1}^{R}\sum_{t=1}^{T_r} d(r,t,i)b'_i(k),\ k = 1,\ldots,K$$

where $b_i(k)$ is multiplied on both sides. Solving for $b_i(k)$, we obtain the reestimation formula:

$$b_i(k) = \frac{\displaystyle\sum_{r=1}^{R}\sum_{\substack{t=1 \\ s.t.x_{r,t}=k}}^{T_r}\sum_{s} p\left(s\,|X,\Lambda'\right)\left(C(s) - O(\Lambda')\right)\gamma_{i,r,s_r}(t) + b'_i(k)\sum_{r=1}^{R}\sum_{t=1}^{T_r} d(r,t,i)}{\displaystyle\sum_{r=1}^{R}\sum_{t=1}^{T_r}\sum_{s} p\left(s\,|X,\Lambda'\right)\left(C(s) - O(\Lambda')\right)\gamma_{i,r,s_r}(t) + \sum_{r=1}^{R}\sum_{t=1}^{T_r} d(r,t,i)} \tag{5.18}$$

We now define

$$D_i = \sum_{r=1}^{R}\sum_{t=1}^{T_r} d(r,t,i) \tag{5.19}$$

$$\Delta\gamma(i,r,t) = \sum_s p\left(s|X,\Lambda'\right)\left(C(s) - O(\Lambda')\right)\gamma_{i,r,s_r}(t) \tag{5.20}$$

and can rewrite (5.18) as

$$b_i(k) = \frac{\displaystyle\sum_{r=1}^{R}\sum_{\substack{t=1\\ s.t.x_{r,t}=k}}^{T_r}\Delta\gamma\left(i,r,t\right) + b_i'(k)D_i}{\displaystyle\sum_{r=1}^{R}\sum_{t=1}^{T_r}\Delta\gamma\left(i,r,t\right) + D_i} \tag{5.21}$$

To optimize transition probabilities $a_{i,j}$, with constraint $\sum_{j=1}^{I}a_{i,j} = 1$, we apply Lagrange multiplier method by constructing

$$W_2(\Lambda;\Lambda') = U_2(\Lambda;\Lambda') + \sum_{i=1}^{I}\lambda_i\left(\sum_{j=1}^{I}a_{i,j} - 1\right) \tag{5.22}$$

Setting $\dfrac{\partial W_2(\Lambda;\Lambda')}{\partial\lambda_i} = 0$ and $\dfrac{\partial W_2(\Lambda;\Lambda')}{\partial a_{i,j}} = 0$ and, $j = 1,\ldots,I$, we have the following $I+1$ equations:

$$\sum_{j=1}^{I}a_{i,j} - 1 = 0$$

$$0 = \lambda_i a_{i,j} + \sum_{r=1}^{R}\sum_{t=1}^{T_r}\overbrace{\sum_s p\left(s|X,\Lambda'\right)\left(C(s) - O(\Lambda')\right)\xi_{i,j,r,s_r}(t)}^{\Delta\xi(i,j,r,t)} + \sum_{r=1}^{R}\sum_{t=1}^{T_r}d(r,t-1,i)a_{i,j}', j = 1,\ldots,I$$

Note that $\sum_{j=1}^{I}\xi_{i,j,r,s_r}(t) = \gamma_{i,r,s_r}(t)$. Solving for $a_{i,j}$, we obtain the reestimation formula with a standard procedure (used for deriving the expectation–maximization estimate of transition probabilities [10]):

$$a_{i,j} = \frac{\displaystyle\sum_{r=1}^{R}\sum_{t=1}^{T_r}\sum_s p\left(s|X,\Lambda'\right)\left(C(s) - O(\Lambda')\right)\xi_{i,j,r,s_r}(t) + a_{i,j}'\sum_{r=1}^{R}\sum_{t=1}^{T_r}d(r,t-1,i)}{\displaystyle\sum_{r=1}^{R}\sum_{t=1}^{T_r}\sum_s p\left(s|X,\Lambda'\right)\left(C(s) - O(\Lambda')\right)\gamma_{i,r,s_r}(t) + \sum_{r=1}^{R}\sum_{t=1}^{T_r}d(r,t-1,i)} \tag{5.23}$$

Now we define

$$\tilde{D}_i = \sum_{r=1}^{R} \sum_{t=1}^{T_r} d(r, t-1, i) \qquad (5.24)$$

$$\Delta\xi(i,j,r,t) = \sum_s p\left(s \,|\, X, \Lambda'\right) \left(C(s) - O(\Lambda')\right) \xi_{i,j,r,s_r}(t) \qquad (5.25)$$

and together with (5.20), we rewrite (5.23) as

$$a_{i,j} = \frac{\displaystyle\sum_{r=1}^{R} \sum_{t=1}^{T_r} \Delta\xi(i,j,r,t) + a'_{i,j}\tilde{D}_i}{\displaystyle\sum_{r=1}^{R} \sum_{t=1}^{T_r} \Delta\gamma(i,r,t) + \tilde{D}_i} \qquad (5.26)$$

The parameter reestimation formulas (5.18) and (5.26) for discrete HMMs are unified across maximum mutual information (MMI), minimum classification error (MCE), and minimum phone error/minimum word error (MPE/MWE). What distinguishes among MMI, MCE, and MPE/MWE is the different weighing term $\Delta\gamma(i,r,t)$ in (5.20) and $\Delta\xi(i,j,r,t)$ in (5.25) due to the different $C(s)$ contained in the unified objective function. Details for computing $\Delta\gamma(i,r,t)$ for MMI, and MCE, and MPE/MWE can be found in Chapter 6.

5.2   ESTIMATION FORMULAS FOR CDHMM

For CDHMMs, $X = X_1, \dots, X_R$, are continuous random variables. The previous objective functions for discrete HMMs still hold, except that χ is a continuous variable and hence the summation over domain χ is changed to integration over χ. Thus, we have the objective function:

$$V(\Lambda; \Lambda') = \sum_s \sum_q \int_\chi f(\chi, q, s, \Lambda') \log f(\chi, q, s, \Lambda) \mathrm{d}\chi \qquad (5.27)$$

where the integrand $f(\chi, q, s, \Lambda)$ is defined by

$$F(\Lambda; \Lambda') = \sum_s \sum_q \int_\chi f(\chi, q, s, \Lambda) \mathrm{d}\chi \qquad (5.28)$$

Correspondingly, we have

$$F(\Lambda;\Lambda') = \sum_{s}\sum_{q}[C(s) - O(\Lambda')]p(s)p(X,q|s,\Lambda) + D$$

$$= \sum_{s}\sum_{q}\int_{\chi}[\Gamma(\Lambda') + d(s)]\,p(\chi,q|s,\Lambda)\mathrm{d}\chi \qquad (5.29)$$

where

$$f(\chi,q,s,\Lambda) = \big[\Gamma(\Lambda') + d(s)\big]p(\chi,q|s,\Lambda) \qquad (5.30)$$

and

$$\Gamma(\Lambda') = \delta(\chi,X)p(s)\big[C(s) - O(\Lambda')\big] \qquad (5.31)$$

with $\delta(\chi, X)$ in (5.31) being the Dirac delta function. After a similar derivation to that in the preceding section, it can be shown that the transition probability estimation formula (5.26) stays the same as for the discrete HMM. But for the emitting probability, (5.16) is changed to

$$U_1(\Lambda;\Lambda') = \sum_{r=1}^{R}\sum_{t=1}^{T_r}\sum_{i=1}^{I}\sum_{s}p\left(s|X,\Lambda'\right)\left(C(s) - O(\Lambda')\right)\gamma_{i,r,s_r}(t)\log b_i(x_{r,t})$$

$$+ \sum_{r=1}^{R}\sum_{t=1}^{T_r}\sum_{i=1}^{I}\mathrm{d}(r,t,i)\int_{\chi_{r,t}}b_i'(\chi_{r,t})\log b_i(\chi_{r,t})\mathrm{d}\chi_{r,t} \qquad (5.32)$$

As the most common member of the CDHMM, we use a Gaussian HMM to derive its parameters' GT formulas (the results for CDHMMs with exponential-family emitting probabilities can be derived similarly). For the Gaussian HMM, $b_i(x_{r,t})$ in (5.32) is a Gaussian distribution:

$$b_i(x_{r,t}) \propto \frac{1}{|\Sigma_i|^{1/2}}\exp\left[-\frac{1}{2}(x_{r,t} - \mu_i)^{\mathrm{T}}\Sigma_i^{-1}(x_{r,t} - \mu_i)\right].$$

where $\{\mu_i, \Sigma_i\}$, $i = 1, 2, \ldots, I$ are the Gaussian mean vectors and covariance matrices.

To solve for μ_i and Σ_i based on (5.32), for the Gaussian at HMM's state i, we set

$$\frac{\partial U_1(\Lambda;\Lambda')}{\partial \mu_i} = 0; \quad \text{and} \quad \frac{\partial U_1(\Lambda;\Lambda')}{\partial \Sigma_i} = 0.$$

This gives:

$$
0 = \sum_{r=1}^{R} \sum_{t=1}^{T_r} \sum_{s} \overbrace{p\left(s|X,\Lambda'\right)\left(C(s) - O(\Lambda')\right)\gamma_{i,r,s_r}(t)}^{\Delta\gamma(i,r,t)} \Sigma_i^{-1}(x_{r,t} - \mu_i)
$$
$$
+ \sum_{r=1}^{R} \sum_{t=1}^{T_r} d(r,t,i)\Sigma_i^{-1} \int_{\chi_{r,t}} b_i'(\chi_{r,t})(\chi_{r,t} - \mu_i)d\chi_{r,t}
\tag{5.33}
$$

$$
0 = \sum_{r=1}^{R} \sum_{t=1}^{T_r} \sum_{s} \overbrace{p\left(s|X,\Lambda'\right)\left(C(s) - O(\Lambda')\right)\gamma_{i,r,s_r}(t)}^{\Delta\gamma(i,r,t)} \left[\Sigma_i^{-1} - \Sigma_i^{-1}(x_{r,t} - \mu_i)(x_{r,t} - \mu_i)^{\mathrm{T}}\Sigma_i^{-1}\right]
$$
$$
+ \sum_{r=1}^{R} \sum_{t=1}^{T_r} d(r,t,i) \int_{\chi_{r,t}} b_i'(\chi_{r,t}) \left[\Sigma_i^{-1} - \Sigma_i^{-1}(\chi_{r,t} - \mu_i)(\chi_{r,t} - \mu_i)^{\mathrm{T}}\Sigma_i^{-1}\right] d\chi_{r,t}
\tag{5.34}
$$

For the Gaussian distribution $b_i'(\chi_{r,t}) = p(\chi_{r,t}|q_{r,t} = i; \Lambda')$, we have

$$
\int_{\chi_{r,t}} b_i'(\chi_{r,t})d\chi_{r,t} = 1,
$$

$$
\int_{\chi_{r,t}} \chi_{r,t} \cdot b_i'(\chi_{r,t})d\chi_{r,t} = \mu_i',
$$

$$
\int_{\chi_{r,t}} (\chi_{r,t} - \mu_i')(\chi_{r,t} - \mu_i')^{\mathrm{T}} \cdot b_i'(\chi_{r,t})d\chi_{r,t} = \Sigma_i'.
$$

Hence, the integrals in (5.33) and (5.34) give closed-form results. Next, we multiply both sides of (5.33) by Σ_i at the left end, and multiply both sides of (5.34) by Σ_i at both left and right ends, respectively. Finally, solving μ_i and Σ_i gives the GT formulas of

$$
\mu_i = \frac{\displaystyle\sum_{r=1}^{R} \sum_{t=1}^{T_r} \Delta\gamma(i,r,t)x_t + D_i\mu_i'}{\displaystyle\sum_{r=1}^{R} \sum_{t=1}^{T_r} \Delta\gamma(i,r,t) + D_i}
\tag{5.35}
$$

$$\Sigma_i = \frac{\sum_{r=1}^{R} \sum_{t=1}^{T_r} \left[\Delta\gamma(i,r,t)(x_t - \mu_i)(x_t - \mu_i)^{\mathrm{T}} \right] + D_i\Sigma_i' + D_i(\mu_i - \mu_i')(\mu_i - \mu_i')^{\mathrm{T}}}{\sum_{r=1}^{R} \sum_{t=1}^{T_r} \Delta\gamma(i,r,t) + D_i} \tag{5.36}$$

where $\Delta\gamma(i, r, t)$ is defined in (5.20) and D_i defined in (5.19).

Just as for the discrete HMM presented in the preceding section, (5.35) and (5.36) are unified parameter estimation formulas for MMI, MCE, and MPE/MWE. And $\Delta\gamma(i, r, t)$ in (5.35) and (5.36) as defined in (5.20) is in the same way as for the discrete distribution case — differing in $C(s)$ for MMI, and MCE, and MPE/MWE, respectively, as will be detailed in Chapter 6.

5.3 RELATIONSHIP WITH GRADIENT-BASED METHODS

The relation between the GT method and gradient-based search method has been studied in the literature (e.g., [3, 46]). It can be shown that, with carefully selected, parameter-dependent step sizes, these two methods can be made identical. However, as was shown in [3] for MMI, the GT-based updating formula (5.35) is best viewed not as a simple gradient ascent but as an approximation to a quadratic Newton update; that is, it can be formulated as a gradient ascent with the step size that approximates inverse Hessian matrix H of the objective function. In the following, we will show the similar relationship for all MMI, MCE, and MPE/MWE enabled by our unifying framework.

Let us use *mean* vector estimation as an example. The gradient of $O(\Lambda)$ with respect to μ_i can be shown as:

$$\nabla_{\mu_i} O(\Lambda)|_{\Lambda=\Lambda'} = \Sigma_i'^{-1} \sum_{r=1}^{R} \sum_{t=1}^{T_r} \Delta\gamma(i,r,t)(x_t - \mu_i') \tag{5.37}$$

On the other hand, we can rewrite the GT formula of (5.35) into the following equivalent form

$$\mu_i = \mu_i' + \frac{1}{\sum_{r=1}^{R} \sum_{t=1}^{T_r} \Delta\gamma(i,r,t) + D_i} \cdot \sum_{r=1}^{R} \sum_{t=1}^{T_r} \Delta\gamma(i,r,t)(x_t - \mu_i')$$

$$= \mu_i' + \frac{1}{\sum_{r=1}^{R} \sum_{t=1}^{T_r} \Delta\gamma(i,r,t) + D_i} \Sigma_i' \cdot \nabla_{\mu_i} O(\Lambda)|_{\Lambda=\Lambda'} \tag{5.38}$$

Consider the quadratic Newton update, where the Hessian H_i for μ_i can be approximated by the following equation after dropping the dependency of μ_i with $\Delta\gamma(i,r,t)$.

$$H_i = \nabla^2_{\mu_i} O(\Lambda)|_{\Lambda=\Lambda'} \approx -\Sigma_i'^{-1} \sum_{r=1}^{R} \sum_{t=1}^{T_r} \Delta\gamma(i,r,t)$$

Therefore, the updating formula of GT in (5.35) can be further rewritten to

$$\mu_i \approx \mu_i' - \underbrace{\frac{\displaystyle\sum_{r=1}^{R}\sum_{t=1}^{T_r}\Delta\gamma(i,r,t)}{\displaystyle\sum_{r=1}^{R}\sum_{t=1}^{T_r}\Delta\gamma(i,r,t) + D_i}}_{\varepsilon_i} H_i^{-1}\nabla_{\mu_i}O(\Lambda)|_{\Lambda=\Lambda'} = \mu_i' + \varepsilon_i\nabla_{\mu_i}O(\Lambda)|_{\Lambda=\Lambda'} \quad (5.39)$$

Compared with simple gradient ascent optimization that the model parameters Λ is updated by

$$\Lambda = \Lambda' + \varepsilon \cdot \nabla O(\Lambda)|_{\Lambda=\Lambda'}$$

where the step size ε is a single, global constant independent of the parameters. Equation (5.39) can be viewed either as a generalization of the simple gradient ascent where the global step size ε is replaced by the Gaussian-dependent step size ε_i, or as a generalization of the quadratic Newton update $\mu_i = \mu_i' - \alpha \cdot H_i^{-1}\nabla_{\mu_i}O(\Lambda)|_{\Lambda=\Lambda'}$. Thus, the GT formula of (5.35) leads to more rapid convergence than the simple gradient-based search.

5.4   SETTING CONSTANT D FOR GT-BASED OPTIMIZATION

Based on Jensen's inequality, the theoretical basis for setting D_i is the requirement described in (5.2). That is, d(s) must be sufficiently large to ensure that for any string s and any observation sequence χ, $\Gamma(\Lambda') + d(s) > 0$, where $\Gamma(\Lambda') = \delta(\chi, X)p(s)[C(s) - O(\Lambda')]$ from (5.3). However, for the CDHMM, $\delta(\chi, X)$ becomes the Dirac delta function, which is unbounded at the Center point. That is, $\delta(\chi, X) = +\infty$ when $\chi = X$. Therefore, for the string s that gives $C(s) - O(\Lambda') < 0$, $\Gamma(\Lambda')|_{\chi=X} = -\infty$. Under this condition, it is impossible to find a bounded d(s) that ensures $\Gamma(\Lambda') + d(s) > 0$ and hence Jensen's inequality may not apply. (Note that the discrete HMM does not encounter such a difficulty because $\delta(\chi, X)$ takes final values of 0 or 1.)

The above difficulty for CDHMMs can be overcome if it can be shown that there exists a sufficiently large but still bounded constant D so that $V(\Lambda; \Lambda')$ of (5.27), with the integrand defined by (5.30) is still a valid auxiliary function of $F(\Lambda; \Lambda')$; that is, an increase in the value of $V(\Lambda; \Lambda')$

can guarantee the increase in the value of $F(\Lambda; \Lambda')$. Such a proof has indeed been developed in the recent work of Axelrod et al. [3], which we will outline in the later part of this section. (Because $V(\Lambda; \Lambda')$ is still a valid auxiliary function, all derivations from (5.27) to (5.36) are valid.)

Given sufficiently large D_i, the convergence of the model estimation formulas, that is, (5.21), (5.26), (5.35), and (5.36), can be proved. However, the value of D_i that guarantees convergence is usually too large to obtain a reasonable convergence speed. Before further research advance to lower the value of D_i, in practice, D_i is often empirically set to achieve compromised training performance.

Empirical setting of D_i has been extensively studied since GT/EBW was proposed. In the early days, only one global constant D was used for all parameters [14, 34]. Later research discovered on the empirical basis that for CDHMM, a useful lower bound on (nonglobal) D_i is the value satisfying the constraint that the newly estimated variances remain positive [35]. In Refs. [50, 51], this constraint was further explored, leading to quadratic inequalities with which the lower bound of D_i can be solved. Most recently, in [46], constant D_i was further bounded by an extra condition that the denominators in the reestimation formula remain nonsingular.

In [52], use of Gaussian-specific D_i was reported to give further improved convergence speed. For MMI, the Gaussian-specific constant D_i was set empirically to be the maximum of (i) twice the value necessary to ensure positive variance, that is, $2 \cdot D_{\min}$; and (ii) a global constant E multiplied by the denominator occupancy; for example, $E \cdot \gamma_i^{\text{den}}$. Specifically, for MMI in the work of Woodland and Povey [52], $\gamma_i^{\text{den}} = \sum_{r=1}^R \sum_{t=1}^T \gamma_{i,r}^{\text{den}}(t) = \sum_{r=1}^R \sum_{t=1}^{T_r} \sum_{s_r} p(s_r|X_r, \Lambda')\gamma_{i,r,s_r}(t)$. However, the $\Delta\gamma(i,r,t)$ in the unified reestimation formulas (5.21), (5.26), (5.35), and (5.36) is different from the classical form in [52] by a constant factor and therefore the setting of D_i should be adjusted accordingly. This issue is discussed in details in Section 6.1. For MPE reported in [38–40], the empirical setting of D_i was the same as MMI, that is, $D_i = \max\{2 \cdot D_{\min}, E \cdot \gamma_i^{\text{den}}\}$ except that the computation of the denominator occupancy became: $\gamma_i^{\text{den}} = \sum_{r=1}^R \sum_{t=1}^{T_r} \max(0, -\Delta\gamma(i, r, t))$. Moreover, the obtained new parameters were smoothed with the ML estimate of parameters (which was called I-smoothing).

For MCE, in our previous experimental work [20, 58], we developed the empirical setting of γ_i^{den} as $\sum_{r=1}^R \sum_{t=1}^{T_r} p(S_r|X_r, \Lambda')\gamma_{i,r,s_r}(t)$. It was based on the consideration that MCE and MMI are equivalent in the special case of having one utterance in the training set and hence the parameter estimation formulas of them should be identical in this special case. We tested this setting and obtained strong results as reported in [20, 58]. Further discussions and comparisons of different settings of empirical D_i can be found in [14, 20, 34, 35, 40, 46, 51, 52].

5.4.1. Existence Proof of Finite D in GT Updates for CDHMM

As discussed earlier, optimization based on Jensen's inequality cannot be applied directly to Gaussian CDHMM because the value D in the GT update formulas (5.35) and (5.36) may be infinite, making the algorithm's convergence infinitely slow. In this section, we follow the insight provided in [3] to prove that there exist finite values of D that make the GT update formulas (5.35) and (5.36) practical for all MMI, MCE, and MPE/MWE.

To proceed, we substitute (5.30) into (5.27) and obtain

$$V(\Lambda;\Lambda') = \sum_q \sum_s \int_\chi \left[\Gamma(\Lambda') + \mathrm{d}(s)\right] p\left(\chi,q|s,\Lambda'\right) \log p\left(\chi,q|s,\Lambda\right) + \text{Const.} \qquad (5.40)$$

We prove below that for a CDHMM, given a sufficiently large but bounded (i.e., finite) constant D,

$$F(\Lambda;\Lambda') - F(\Lambda';\Lambda') \geq V(\Lambda;\Lambda') - V(\Lambda';\Lambda') \qquad (5.41)$$

First, we define

$$\Delta_D = \left[F(\Lambda;\Lambda') - F(\Lambda';\Lambda')\right] - \left[V(\Lambda;\Lambda') - V(\Lambda';\Lambda')\right] \qquad (5.42)$$

and will show that $\Delta_D \geq 0$ for any parameter set Λ. Substituting (5.29) and (5.40) into (5.42), we obtain

$$\Delta_D = \left[F(\Lambda;\Lambda') - F(\Lambda';\Lambda')\right] - \left[V(\Lambda;\Lambda') - V(\Lambda';\Lambda')\right]$$

$$= \sum_q \sum_s \int_\chi \left[\Gamma(\Lambda') + \mathrm{d}(s)\right] \left[p\left(\chi,q|s,\Lambda\right) - p(\chi,q|s,\Lambda')\right] \mathrm{d}\chi$$

$$\quad - \sum_q \sum_s \int_\chi \left[\Gamma(\Lambda') + \mathrm{d}(s)\right] p\left(\chi,q|s,\Lambda'\right) \left[\log p\left(\chi,q|s,\Lambda\right) - \log p\left(\chi,q|s,\Lambda'\right)\right] \mathrm{d}\chi \qquad (5.43)$$

$$= \sum_q \sum_s \int_\chi \left[\Gamma(\Lambda') + \mathrm{d}(s)\right] p\left(\chi,q|s,\Lambda'\right) \left[\frac{p\left(\chi,q|s,\Lambda\right)}{p\left(\chi,q|s,\Lambda'\right)} - 1 - \log\frac{p\left(\chi,q|s,\Lambda\right)}{p\left(\chi,q|s,\Lambda'\right)}\right] \mathrm{d}\chi$$

$$= \sum_q \sum_s \int_\chi \left[\Gamma(\Lambda') + \mathrm{d}(s)\right] p\left(\chi,q|s,\Lambda'\right) \mathrm{H}\left(\chi,q,s,\Lambda,\Lambda'\right) \mathrm{d}\chi$$

where

$$\mathrm{H}\left(\chi,q,s,\Lambda,\Lambda'\right) = \left[\frac{p\left(\chi,q|s,\Lambda\right)}{p\left(\chi,q|s,\Lambda'\right)} - 1\right] - \log\left[\left[\frac{p\left(\chi,q|s,\Lambda\right)}{p\left(\chi,q|s,\Lambda'\right)} - 1\right] + 1\right]$$

Then, we need to show that there exists a bounded $d(s)$ that ensures the summand of Δ_D in (5.43) be nonnegative. To proceed, we expand the summand to

$$\int_\chi \left[\Gamma(\Lambda') + d(s)\right] p\left(\chi, q \,|\, s, \Lambda'\right) H\left(\chi, q, s, \Lambda, \Lambda'\right) d\chi$$

$$= p(s)\left[C(s) - O(\Lambda')\right] p\left(X, q \,|\, s, \Lambda'\right) H(X, q, s, \Lambda, \Lambda') + d(s) \int_\chi p(\chi, q \,|\, s, \Lambda') H(\chi, q, s, \Lambda, \Lambda') d\chi \qquad (5.44)$$

We now use the following key theorem from [3]: If $f(X, \Lambda)$ is nonnegative and analytic for $X \in \chi$ and $\Lambda \in \Omega$, where χ and Ω are the data space and model space, respectively, then there is a Λ-independent constant $K > 0$ such that

$$\int_\chi f(\chi, \Lambda) d\chi \geq K f(X, \Lambda) \qquad (5.45)$$

for any valid model Λ. (Readers are referred to [3] for a rigorous proof.)

Define $f(X, \Lambda) = p(X, q \,|\, s, \Lambda') H(X, q, s, \Lambda, \Lambda')$. Here, $f(X, \Lambda)$ is nonnegative and analytic because both $p(X, q \,|\, s, \Lambda')$ and $H(X, q, s, \Lambda, \Lambda')$ are nonnegative and analytic (for CDHMM). Using (5.45), we have

$$\int_\chi p\left(\chi, q \,|\, s, \Lambda'\right) H\left(\chi, q, s, \Lambda, \Lambda'\right) d\chi \geq K p\left(X, q \,|\, s, \Lambda'\right) H\left(X, q, s, \Lambda, \Lambda'\right) \qquad (5.46)$$

Now we construct nonnegative d(s) as follows:

$$d(s) = \begin{cases} 0 & \text{if } C(s) \geq O(\Lambda') \\ \dfrac{1}{K} p(s) \left(O(\Lambda') - C(s)\right) & \text{if } C(s) < O(\Lambda') \end{cases} \qquad (5.47)$$

Then, (5.46) becomes

$$d(s) \int_\chi p\left(\chi, q \,|\, s, \Lambda'\right) H\left(\chi, q, s, \Lambda, \Lambda'\right) d\chi > -p(s) \left[C(s) - O(\Lambda')\right] p\left(X, q \,|\, s, \Lambda'\right) H\left(X, q, s, \Lambda, \Lambda'\right)$$

This proves that the summand of Δ_D, $\int_\chi [\Gamma(\Lambda') + d(s)] p(\chi, q \,|\, s, \Lambda') H(X, q, s, \Lambda, \Lambda') d\chi$, is nonnegative for any s (according to (5.44)), and therefore $\Delta_D \geq 0$.

CHAPTER 6

Practical Implementation of Discriminative Learning

The basic development of growth transformation (GT)-based optimization for the hidden Markov model (HMM) and of the unified objective function has been presented in earlier chapters, where several practical considerations and implementation issues are left to this chapter.

6.1 COMPUTING $\Delta\gamma(i,r,t)$ IN GROWTH-TRANSFORM FORMULAS

In (5.20) computing $\Delta\gamma(i,r,t)$ involves summation over all possible superstring label sequences $s = s_1, \ldots, s_R$. The number of training tokens (sentence strings), R, is usually very large. Hence, the summation over s needs to be decomposed and simplified. To proceed, we use the notations of $s' = s_1, \ldots, s_{r-1}, s'' = s_{r+1}, \ldots, s_R, X' = X_1, \ldots, X_{r-1}$, and $X'' = X_{r+1}, \ldots, X_R$. Then, from (5.20), we have,

$$\Delta\gamma(i,r,t) = \sum_{s'}\sum_{s_r}\sum_{s''} p\left(s', s_r, s'' | X', X_r, X''; \Lambda'\right) \left(C(s', s_r, s'') - O(\Lambda')\right) \gamma_{i,r,s_r}(t)$$

$$= \sum_{s_r} p\left(s_r | X_r, \Lambda'\right) \underbrace{\left[\sum_{s'}\sum_{s''} p\left(s', s'' | X', X''; \Lambda'\right) \left(C(s', s_r\ s'') - O(\Lambda')\right)\right]}_{\Psi} \gamma_{i,r,s_r}(t)$$

$$(6.1)$$

where factor Ψ is the average deviation of the accuracy count for the given string s_r. The remaining steps in simplifying the computation of $\Delta\gamma(i,r,t)$ will be separate for maximum mutual information (MMI) and minimum classification error/minimum phone error/minimum word error (MCE/MPE/MWE) because the parameter-independent accuracy count function $C(s)$ for them takes the product and summation form, respectively (as shown in Table 3.1).

6.1.1 Product Form of $C(s)$ (for MMI)

For MMI, we have $C(s) = C(s_1, \ldots, s_R) = \prod_{r=1}^{R} C(s_r) = \prod_{r=1}^{R} \delta(s_r, S_r)$ in a product form. Using $C(s', s_r, s'') = C(s_r) \cdot C(s', s'')$, we simplify factor Ψ in (6.1) to

$$\Psi = \sum_{s'} \sum_{s''} p\left(s', s'' \mid X', X''; \Lambda'\right) \left(C(s', s_r, s'') - O(\Lambda')\right)$$

$$= C(s_r) \cdot \sum_{s'} \sum_{s''} p\left(s', s'' \mid X', X''; \Lambda'\right) C(s', s'') - O(\Lambda')$$

$$= O(\Lambda') \left(\frac{C(s_r) \cdot \sum_{s'} \sum_{s''} p(s', s'' \mid X', X''; \Lambda') C(s', s'')}{O(\Lambda')} - 1 \right) \tag{6.2}$$

The idea behind the above steps is to make use of the product form of the $C(s)$ function for canceling out common factors in both $O(\Lambda')$ and $C(s)$ functions. To proceed, we now factorize $O(\Lambda')$ as follows:

$$O(\Lambda') = \frac{\sum_{s'} \sum_{s_r} \sum_{s''} \left[p\left(s', s_r, s'', X', X_r, X'' \mid \Lambda'\right) C(s', s_r, s'')\right]}{\sum_{s'} \sum_{s_r} \sum_{s''} p\left(s', s_r, s'', X', X_r, X'' \mid \Lambda'\right)}$$

$$= \frac{\left[\sum_{s_r} p(s_r, X_r \mid \Lambda') C(s_r)\right] \left[\sum_{s'} \sum_{s''} p\left(s', s'', X', X'' \mid \Lambda'\right) C(s', s'')\right]}{\left[p(X_r \mid \Lambda')\right]\left[p(X', X'' \mid \Lambda')\right]}$$

$$= \sum_{s_r} \left[p\left(s_r \mid X_r, \Lambda'\right) C(s_r)\right] \cdot \sum_{s'} \sum_{s''} p\left(s', s'' \mid X', X''; \Lambda'\right) C(s', s'')$$

$$= p\left(S_r \mid X_r, \Lambda'\right) \cdot \sum_{s'} \sum_{s''} p\left(s', s'' \mid X', X''; \Lambda'\right) C(s', s'')$$

where the last step uses $C(s_r) = \delta(s_r, S_r)$. Substituting this to (6.2) then gives the simplification of

$$\Psi = O(\Lambda') \left(\frac{C(s_r) \cdot \sum_{s'} \sum_{s''} p\left(s', s'' \mid X', X''; \Lambda'\right) C(s', s'')}{p(S_r \mid X_r, \Lambda') \cdot \sum_{s'} \sum_{s''} p\left(s', s'' \mid X', X'', \Lambda'\right) C(s', s'')} - 1 \right)$$

$$= O(\Lambda') \left(\frac{C(s_r)}{p\left(S_r \mid X_r, \Lambda'\right)} - 1 \right) \tag{6.3}$$

Substituting (6.3) to (6.1) and using $C(s_r) = \delta(s_r, S_r)$ again for MMI, we obtain

$$\Delta\gamma(i,r,t) = O(\Lambda')\sum_{s_r} p\left(s_r\middle|X_r,\Lambda'\right)\left(\frac{C(s_r)}{p\left(S_r\middle|X_r,\Lambda'\right)}-1\right)\gamma_{i,r,s_r}(t)$$

$$= O(\Lambda')\sum_{s_r,\,s_r\neq S_r} p\left(s_r\middle|X_r,\Lambda'\right)\left(\frac{C(s_r)}{p\left(S_r\middle|X_r,\Lambda'\right)}-1\right)\gamma_{i,r,s_r}(t)$$

$$+ O(\Lambda')p\left(S_r\middle|X_r,\Lambda'\right)\left(\frac{C(S_r)}{p\left(S_r\middle|X_r,\Lambda'\right)}-1\right)\gamma_{i,r,S_r}(t)$$

$$= -O(\Lambda')\sum_{s_r,\,s_r\neq S_r} p\left(s_r\middle|X_r,\Lambda'\right)\gamma_{i,r,s_r}(t) + O(\Lambda')\left(1 - p(S_r|X_r,\Lambda')\right)\gamma_{i,r,S_r}(t)$$

$$= O(\Lambda')\left[\gamma_{i,r,S_r}(t) - \sum_{s_r} p\left(s_r\middle|X_r,\Lambda'\right)\gamma_{i,r,s_r}(t)\right] \qquad (6.4)$$

In the reestimation formulas (5.35) and (5.36), if we divide both the numerator and denominator by $O(\Lambda')$, $\Delta\gamma(i,r,t)$ in (6.4) can take a simplified form of

$$\Delta\tilde{\gamma}(i,r,t) = \left[\gamma_{i,r,S_r}(t) - \sum_{s_r} p\left(s_r\middle|X_r,\Lambda'\right)\gamma_{i,r,s_r}(t)\right] = \gamma_{i,r}^{\text{num}}(t) - \gamma_{i,r}^{\text{den}}(t) \qquad (6.5)$$

The corresponding constant D_i in the reestimation formulas (5.35) and (5.36) then becomes

$$\tilde{D}_i = D_i/O(\Lambda') \qquad (6.6)$$

Substituting this into (5.35) and (5.36), we have the GT formulas for MMI

$$\mu_i = \frac{\sum_{r=1}^{R}\sum_{t=1}^{T_r}\left[\gamma_{i,r}^{\text{num}}(t) - \gamma_{i,r}^{\text{den}}(t)\right]x_t + \tilde{D}_i\mu_i'}{\sum_{r=1}^{R}\sum_{t=1}^{T_r}\left[\gamma_{i,r}^{\text{num}}(t) - \gamma_{i,r}^{\text{den}}(t)\right] + \tilde{D}_i} \qquad (6.7)$$

$$\Sigma_i = \frac{\sum_{r=1}^{R}\sum_{t=1}^{T_r}\left[\gamma_{i,r}^{\text{num}}(t) - \gamma_{i,r}^{\text{den}}(t)\right](x_t - \mu_i)(x_t - \mu_i)^{\text{T}} + \tilde{D}_i\Sigma_i' + \tilde{D}_i(\mu_i - \mu_i')(\mu_i - \mu_i')^{\text{T}}}{\sum_{r=1}^{R}\sum_{t=1}^{T_r}\left[\gamma_{i,r}^{\text{num}}(t) - \gamma_{i,r}^{\text{den}}(t)\right] + \tilde{D}_i} \qquad (6.8)$$

This gives the classical GT/EBW-based MMI reestimation formulas described in [34, 52].

Equation (6.4) or (6.5) gives an N-best-string-based solution to computing $\Delta\gamma(i,r,t)$. This is illustrated by the string-level summation over s_r (i.e., the label sequence for token r, including both correct and incorrect strings). For N-best string-level discriminative training, the summation over s_r in (6.4) or (6.5) amounts to going through all N-best string hypotheses and is computationally inexpensive when N is relatively small (e.g., N in the order of thousands as typical for most N-best experiments).

When a lattice instead of an explicit N-best list is provided for competing hypotheses in discriminative training, in theory, (6.4) or (6.5) can be applied just as for the N-best string based solution already discussed. This is because a lattice is nothing more than a compact representation of N-best strings. However, because N in this equivalent "N-best list" would be huge (in the order of billions or higher [57]), more efficient techniques for dealing with the summation over s_r in computing (6.4) or (6.5) will be needed. Readers are referred to Section 6.2 for details of such computation.

6.1.2. Summation Form of $C(s)$ (MCE and MPE/MWE)

Different from MMI, for MCE and MPE/MWE, we have $C(s) = C(s_1, \ldots, s_R) = \sum_{r=1}^{R} C(s_r)$, or $C(s', s_r, s'') = C(s_r) + C(s', s'')$. That is, the C function is in a summation instead of a product form. This changes the simplification steps for factor Ψ of (6.1) as follows:

$$
\Psi = \sum_{s'} \sum_{s''} p\left(s', s'' \mid X', X''; \Lambda'\right) \left(C(s', s_r, s'') - O(\Lambda') \right)
$$

$$
= \sum_{s'} \sum_{s''} p_{\Lambda'}\left(s', s'' \mid X', X''\right) C(s_r) + \sum_{s'} \sum_{s''} p\left(s', s'' \mid X', X''; \Lambda'\right) C(s', s'')
$$

$$
- \sum_{s'} \sum_{s''} p\left(s', s'' \mid X', X''; \Lambda'\right) O(\Lambda') = C(s_r) + \sum_{s'} \sum_{s''} p\left(s', s'' \mid X', X''; \Lambda'\right) C(s', s'') - O(\Lambda')
$$

$$(6.9)$$

The idea behind the above steps is to make use of the summation form of the $C(s)$ function for subtracting out the common terms in the $O(\Lambda')$ function. To achieve this, we decompose $O(\Lambda')$, based on its original nonrational form (3.20) or (3.22), (3.23) as follows:

$$
O(\Lambda') = \sum_{i=1}^{R} \frac{\sum_{s_i} p\left(X_i, s_i \mid \Lambda'\right) C(s_i)}{\sum_{s_i} p\left(X_i, s_i \mid \Lambda'\right)}
$$

$$
= \frac{\sum_{s_r} p\left(X_r, s_r \mid \Lambda'\right) C(s_r)}{\sum_{s_r} p\left(X_r, s_r \mid \Lambda'\right)} + \sum_{i=1, i \neq r}^{R} \frac{\sum_{s_i} p\left(X_i, s_i \mid \Lambda'\right) C(s_i)}{\sum_{s_i} p\left(X_i, s_i \mid \Lambda'\right)}
$$

$$
= \frac{\sum_{s_r} p\left(X_r, s_r \mid \Lambda'\right) C(s_r)}{\sum_{s_r} p\left(X_r, s_r \mid \Lambda'\right)} + \frac{\sum_{s', s''} p\left(s', s'', X', X'' \mid \Lambda'\right) C(s', s'')}{\sum_{s', s''} p\left(s', s'', X', X'' \mid \Lambda'\right)}
$$

$$
= \frac{\sum_{s_r} p\left(X_r, s_r \mid \Lambda'\right) C(s_r)}{\sum_{s_r} p\left(X_r, s_r \mid \Lambda'\right)} + \sum_{s', s''} p\left(s', s'' \mid X', X'' \mid \Lambda'\right) C(s', s'')
$$

The second term above cancels out the same term in (6.9), leading to the simplification of

$$\Psi = C(s_r) - \frac{\sum_{s_r} p\left(s_r, X_r | \Lambda'\right) C(s_r)}{\sum_{s_r} p\left(s_r, X_r | \Lambda'\right)} \tag{6.10}$$

Now, substituting (6.10) back to (6.1), we obtain

$$\Delta\gamma(i,r,t) = \sum_{s_r} p\left(s_r | X_r, \Lambda'\right) \left(C(s_r) - \frac{\sum_{s_r} p\left(X_r, s_r | \Lambda'\right) C(s_r)}{\sum_{s_r} p\left(X_r, s_r | \Lambda'\right)} \right) \gamma_{i,r,s_r}(t) \tag{6.11}$$

For MCE that has $C(s_r) = \delta(s_r, S_r)$, the above equation can be further simplified to

$$\Delta\gamma(i,r,t) = p\left(S_r | X_r, \Lambda'\right) \left[\gamma_{i,r,S_r}(t) - \sum_{s_r} p\left(s_r | X_r, \Lambda'\right) \gamma_{i,r,s_r}(t) \right] \tag{6.12}$$

Again, if a lattice instead of an N-best list is provided for discriminative learning, a huge number of terms in the summation over s_r in (6.11) would be encountered. To keep the computation manageable, one needs to approximate the computation in (6.11), which we describe below.

6.2   COMPUTING $\Delta\gamma(i,r,t)$ USING LATTICES

A lattice, as illustrated in Figure 6.1, is a compact representation of a large list of strings. It is an acyclic directed graph consisting of a number of nodes (nine in Figure 6.1 as a highly simplified example) and a set of directed arcs each connecting two nodes. In Figure 6.1, each node corresponds to a time stamp and each arc corresponds to a substring unit (e.g., a word of a phone in a sentence). A string in the lattice contains multiple arcs. A typical arc is shown as q in Figure 6.1. Two time stamps, b_q and e_q, are associated with each arc, providing an estimate of the segment boundaries for the substring. For a time slice t within the arc segment q, we have $b_q \leq t \leq e_q$.

We will show below that (6.5) and (6.11) can both be computed efficiently by a forward–backward algorithm. First, given the lattice in Figure 6.1 and s_r as an arbitrary path in that lattice, we will show the occupancy given the entire string s_r can be computed as the occupancy given the local arc q, where arc q belongs to s_r. that is,

$$\gamma_{i,r,q}(t) = \gamma_{i,r,s_r}(t) \text{ when } b_q \leq t \leq e_q \tag{6.13}$$

To see this, let s_r be composed of three substrings: s'_r, q, s'', and correspondingly the observation sequence X_r is composed of three subsequences: X''_r, X_q, X'''. Then the right-hand side of (6.13) can be analyzed as

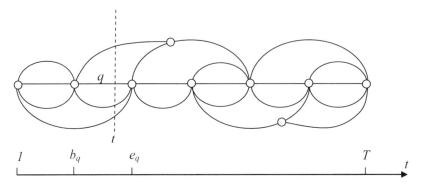

FIGURE 6.1: A graphical illustration of a lattice, where q represents an arc in the lattice and t represents a time slice. The time span of arc q is $b_q \leq t \leq e_q$ and that for the entire lattice is $1 \leq t \leq T$. In this simple example, the total number of arcs (q) is 21, which is substantially lower than the total number of paths (s_r), which can be counted to be as 420. The essence of the decomposition of occupation probability introduced in the text (Eq. (6.13)) is to enable fast computation by reducing the number of terms in summation over s_r to that over q.

$$\gamma_{i,r,s_r}(t:b_q \leq t \leq e_q) = p\left(q_{r,t:b_q \leq t \leq e_q} = i \,\middle|\, X_r, s_r, \Lambda'\right)$$

$$= p\left(q_{r,t:b_q \leq t \leq e_q} = i \,\middle|\, X'_r, X_q, X''_r, s'_r, q, s''_r, \Lambda'\right)$$

$$= p\left(q_{r,t:b_q \leq t \leq e_q} = i \,\middle|\, X_q, q, \Lambda'\right)$$

$$= \gamma_{i,r,q}(t:b_q \leq t \leq e_q)$$

which is the left-hand side of (6.13). The third step holds because the HMM of s_r is formed by concatenating phone-specific HMMs, so that the states in different arcs belong to different HMMs, and are independent of each other, that is, given arc q, its first HMM state $q_{r,b}$ is independent of its preceding state q_{r,b_q-1}.

The essence of (6.13) is to decouple the dependency on the local arc q from the entire string s_r. This enables drastic simplification of the computation in (6.5) and (6.11), which we discuss below for three separate cases.

6.2.1   Computing $\Delta\gamma(i,r,t)$ for MMI Involving Lattices

The principal computation burden in (6.5) is the huge number (N) of summation terms for s_r for the equivalent N-best list of a lattice in the following quantity in (6.5):

$$\Upsilon = \sum_{s_r} p\left(s_r \middle| X_r, \Lambda'\right) \gamma_{i,r,s_r}(t) \tag{6.14}$$

Using (6.13), we can significantly reduce the computation by the following simplification:

$$\Upsilon = \sum_{s_r} p\left(s_r \middle| X_r, \Lambda'\right) \gamma_{i,r,q}(t) = \sum_{q : t \in [b_q, e_q]} \gamma_{i,r,q}(t) \cdot \sum_{s_r : q \in s_r} p\left(s_r \middle| X_r, \Lambda'\right)$$

$$= \sum_{q : t \in [b_q, e_q]} \gamma_{i,r,q}(t) \cdot p\left(q \middle| X_r, \Lambda'\right) = \sum_{q : t \in [b_q, e_q]} \gamma_{i,r,q}(t) \cdot \frac{p\left(q, X_r \middle| \Lambda'\right)}{p\left(X_r \middle| \Lambda'\right)} \tag{6.15}$$

Note that the number of summation terms for q in (6.15) after the approximation is substantially smaller than that for s_r before the approximation. The key quantities in (6.15) can be efficiently computed as follows (proof omitted):

$$p(q, X_r \middle| \Lambda') = \alpha(q)\beta(q) \tag{6.16}$$

$$p\left(X_r \middle| \Lambda'\right) = \sum_{q : q \in \{\text{ending arcs}\}} p\left(q, X_r \middle| \Lambda'\right) = \sum_{q : q \in \{\text{ending arcs}\}} \alpha(q) \tag{6.17}$$

where the "forward" and "backward" probabilities are defined by

$$\alpha(q) \triangleq p\left(q, X'_r(q), X_r(q) \middle| \Lambda'\right) \tag{6.18}$$

$$\beta(q) \triangleq p\left(X''_r(q) \middle| q, \Lambda'\right) \tag{6.19}$$

In (6.18), $X'_r(q)$ denotes the rth training token's partial observation sequence preceding arc q, that is, during $1 \leq t < b_q$. $X_r(q)$ is the observation sequence bounded by arc q with $b_q \leq t \leq e_q$. $X''_r(q)$ in (6.19) denotes the partial observation sequence succeeding arc q, or during $e_q < t \leq T_r$. $\alpha(q)$ is the probability that lattice is at arc q during time $b_q \leq t \leq e_q$, and having generated partial observation $X'_r(q)$ plus $X_r(q)$, that is, $x_{r,1}, \ldots, x_{r,e_q}$. $\beta(q)$ is the probability of generating partial observation $X''_r(q)$ given that the lattice is at arc q at time $t = e_q$.

For each arc q in the lattice, $\alpha(q)$ and $\beta(q)$ can be computed by the following efficient forward and backward recursions, respectively (proofs omitted):

$$\alpha(q) = \sum_{\{p : p \text{ succeeds } q\}} p\left(q \middle| p, \Lambda'\right) p\left(X_r(q) \middle| q, \Lambda'\right) \alpha(p) \tag{6.20}$$

and

$$\beta(q) = \sum_{\{v : v \text{ succeeds } q\}} p\left(v \middle| q, \Lambda'\right) p\left(X_r(v) \middle| v, \Lambda'\right) \beta(v) \tag{6.21}$$

where in (6.20), $\{p: p$ precedes $q\}$ is the collection of all arcs p that directly connects to q in the lattice. Similarly, $\{v: v$ succeeds $q\}$ in (6.21) is the collection of all arcs v that directly connect to q in the lattice. $\alpha(q)$ is initialized at the starting arc q_0 by $\alpha(q_0) = \pi(q_0)p(X_r(q_0)|q_0, \Lambda')$, and $\beta(q)$ initialized at the ending arc q_E by $\beta(q_E) = 1$.

The recursive computation of $\alpha(q)$ and $\beta(q)$ is illustrated in Figure 6.2. There is a direct analogy between this forward and backward probability computation over the sublattice illustrated here and that for the standard HMM over time [10, 43]. In Figure 6.2, the arc q under consideration is analogous to the HMM state occupied at current time frame t in describing the HMM's forward–backward algorithm, the set of arcs $\{p: p$ precedes $q\}$ is analogous to all states in HMM at frame $t - 1$, the set $\{v: v$ succeeds $q\}$ is analogous to all states in HMM at frame $t + 1$. $X_r'(q)$ plays the role of the sequence of observation vectors from 1 to $t - 1$, and $X_r''(q)$ plays the role of the sequence of observation vectors from $t + 1$ to the end. $P(q|p,\Lambda')$ is analogous to the HMM's transition probability (and its value is available from the lattice as the phone or word's "bigram language model" score). $p(X_r(q)|q,\Lambda')$ is analogous to the HMM's emission probability (and its value is available from the lattice as the "acoustic model" score for arc q). Given these analogies, the forward and backward probability computation for (6.20) and (6.21) as illustrated in Figure 6.2 becomes identical to that for the standard HMM (as illustrated in Figures 6.5 and 6.6 of Ref. [43]).

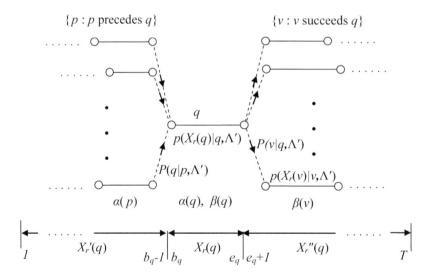

FIGURE 6.2: Illustrations of the sublattice that contains arc q and of the computation of the forward and backward $\alpha(q)$ and $\beta(q)$ based on the sublattice. Each solid line represents an arc in the lattice, and each dashed line represents the direct connection between two arcs (i.e., $b_q - 1 = e_p$).

6.2.2 Computing $\Delta\gamma(i, r, t)$ for MPE/MWE Involving Lattices

We now describe how the computation burden in (6.11) due to the huge number of summation terms over string s_r can be drastically reduced for the MPE/MWE case. It should be pointed out that (6.11) is a unified form for both MCE and MPE/MWE. However, due to the different properties of $C(s_r)$ (i.e., MCE has each term as the Kronecker delta function, but not so for MPE/MWE), the lattice-based computation of (6.11) for MCE and MPE/MWE becomes different.

Consider a particular string token s_r that consists of a sequence of subtokens or substrings. For MCE, $C(s_r) = \delta(s_r, S_r)$, and hence if any of the subtokens is incorrect, the entire token is incorrect also. On the other hand, for MPE, $C(s_r) = A(s_r, S_r)$, which is the raw phone (substring) accuracy count in the sentence string s_r. Therefore, we have a sum of raw phone (substring) accuracy counts of all subtokens; that is, for $s_r = s_{r,1}, \ldots, s_{r,N}$, we have $C(s_r) = \sum_{i=1}^{N} C(s_{r,i})$, where $C(s_{r,i})$ is the raw accuracy count of the subtoken $s_{r,i}$. Readers are referred to [40] for the computation of $C(s_{r,i})$ for subtoken $s_{r,i}$ in the lattice.

In this section, we discuss the lattice-based computation of (6.5) for MPE/MWE. (The lattice-based MCE will be discussed in the next section.)

To proceed, we define

$$\bar{C}_r \triangleq \frac{\sum_{s_r} p\left(X_r, s_r | \Lambda'\right) C(s_r)}{\sum_{s_r} p\left(X_r, s_r | \Lambda'\right)} \tag{6.22}$$

which is the average accuracy count of utterance r, given the observation sequence (X_r) and the lattice that represents all possible strings s_r.

Then, we make use of (6.13) to simplify (6.11) as follows:

$$\Delta\gamma(i, r, t) = \sum_{s_r} p\left(s_r | X_r, \Lambda'\right) \left(C(s_r) - \bar{C}_r\right) \gamma_{i,r,s_r}(t)$$

$$= \sum_{q:t\in[b_q, e_q]} \sum_{s_r:q\in s_r} p\left(s_r | X_r, \Lambda'\right) \left(C(s_r) - \bar{C}_r\right) \gamma_{i,r,q}(t)$$

$$= \sum_{q:t\in[b_q, e_q]} \gamma_{i,r,q}(t) \cdot \left[\sum_{s_r:q\in s_r} p\left(s_r | X_r, \Lambda'\right) C(s_r) - \bar{C}_r \cdot \sum_{s_r:q\in s_r} p\left(s_r | X_r, \Lambda'\right)\right]$$

$$= \sum_{q:t\in[b_q, e_q]} \gamma_{i,r,q}(t) \cdot \left[p\left(q | X_r, \Lambda'\right) \cdot \frac{\sum_{s_r:q\in s_r} p\left(s_r | X_r, \Lambda'\right) C(s_r)}{p\left(q | X_r, \Lambda'\right)} - \bar{C}_r \cdot p\left(q | X_r, \Lambda'\right)\right]$$

$$= \sum_{q:t\in[b_q, e_q]} \gamma_{i,r,q}(t) \cdot p\left(q | X_r, \Lambda'\right) \cdot \left[\bar{C}_r(q) - \bar{C}_r\right] \tag{6.23}$$

where $p(q|X_r, \Lambda') = \sum_{s_r:q \in s_r} p(s_r|X_r, \Lambda') = \dfrac{p(q|X_r, \Lambda')}{p(X_r, \Lambda')}$ is computed in the same way as for (6.16) and (6.17). In (6.23), we define

$$\bar{C}_r(q) = \frac{\displaystyle\sum_{s_r:q \in s_r} p(s_r|X_r, \Lambda') C(s_r)}{p(q|X_r, \Lambda')} = \frac{\displaystyle\sum_{s_r:q \in s_r} p(s_r, X_r|\Lambda') C(s_r)}{\displaystyle\sum_{v_r:q \in v_r} p(v_r, X_r|\Lambda')} \qquad (6.24)$$

which is the average accuracy count of the utterance r, given observation sequence X_r and the sub-lattice that represents all strings s_r containing arc q.

The difficulty of computing $\bar{C}_r(q)$ and $\bar{C}_r$ in (6.23) lies in the very large number of terms in the summation over $s_r: q \in s_r$ and over s_r, respectively. To efficiently compute $\bar{C}_r(q)$ and $\bar{C}_r$, we now further define the following two additional "forward" and "backward" variables for each arc q (following [40]):

$$\varphi(q) \triangleq \frac{\displaystyle\sum_{\{s':s' \text{ precedes } q\}} p(s',q,X'_r(q),X_r(q)|\Lambda') C(s',q)}{\displaystyle\sum_{\{s':s' \text{ precedes } q\}} p(s',q,X'_r(q),X_r(q)|\Lambda')} \qquad (6.25)$$

and

$$\psi(q) \triangleq \frac{\displaystyle\sum_{\{s'':s'' \text{ succeeds } q\}} p(s'',X''_r(q)|q,\Lambda') C(s'')}{\displaystyle\sum_{\{s'':s'' \text{ succeeds } q\}} p(s'',X''_r(q)|q,\Lambda')} \qquad (6.26)$$

In (6.25), $\varphi(q)$ is the weighted average accuracy count of the sublattice that represents all partial paths (s',q) ending inclusively in q, with the partial observation sequence $X'_r(q) \cup X_r(q)$ (i.e., $x_{r,1}, \ldots, x_{r,\text{eq}}$). In (6.26), $\psi(q)$ is the weighted average accuracy count of the sublattice that represents all partial paths s'' that succeeds q, with the partial observation sequence $X'_r(q)$. Figure 6.3 illustrates the sublattice that represents all s_r that contains arc q, together with all the relevant quantities for defining $\varphi(q)$ and $\psi(q)$ based on the sublattice. To show these quantities in defining $\varphi(q)$, we denote the accuracy count as $C(s',q)$ for a given partial path (s',q) encircled by the dotted line to the left of Figure 6.3. We denote the weight associated with this partial path as $p(s', q, X'_r(q), X_r(q)|\Lambda')$. The relevant quantities defining $\psi(q)$ are illustrated to the right of Figure 6.3, including the partial path (s'') that is to the future of arc q, the accuracy count $C(s'')$ associated with this path, and the associated weight of $p(s'', X''_r(q)|\Lambda')$.

We now describe the computation of $\varphi(q)$ as defined in (6.25) and $\psi(q)$ defined in (6.26) efficiently for each arc q in the lattice. For $\varphi(q)$, we use the following efficient "forward" recursion (proof omitted):

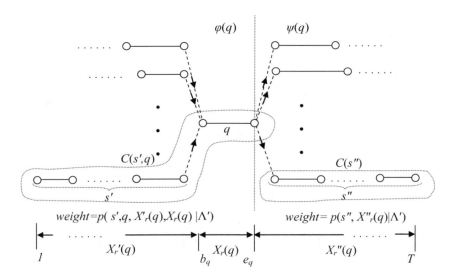

FIGURE 6.3: Illustrations of the sublattice that contains arc q, and of the probability weights that define $\varphi(q)$ of (6.25) and $\psi(q)$ of (6.26) based on the sublattice.

$$\varphi(q) = \frac{\displaystyle\sum_{\{p:p \text{ precedes } q\}} P\big(q\big|p,\Lambda'\big)\,p\big(X_r(q)\big|q,\Lambda'\big)\,\alpha(p)[\varphi(p) + C(q)]}{\displaystyle\sum_{\{p:p \text{ precedes } q\}} P\big(q\big|p,\Lambda'\big)\,p\big(X_r(q)\big|q,\Lambda'\big)\,\alpha(p)}$$

$$= \frac{\displaystyle\sum_{\{p:p \text{ precedes } q\}} P\big(q\big|p,\Lambda'\big)\,\alpha(p)\varphi(p)}{\displaystyle\sum_{\{p:p \text{ precedes } q\}} P\big(q\big|p,\Lambda'\big)\,\alpha(p)} + C(q) \qquad (6.27)$$

where $\varphi(q)$ is initialized for each starting arc q_0 by $\varphi(q_0) = C(q_0)$, which is the raw phone or word accuracy for q_0. For $\psi(q)$, we use the following efficient "backward" recursion (proof omitted):

$$\psi(q) = \frac{\displaystyle\sum_{\{v:v \text{ succeeds } q\}} p\big(X_r(v)\big|v,\Lambda'\big)\,P\big(v\big|q,\Lambda'\big)\,\beta(v)\,[C(v) + \psi(v)]}{\displaystyle\sum_{\{v:v \text{ succeeds } q\}} p\big(X_r(v)\big|v,\Lambda'\big)\,P\big(v\big|q,\Lambda'\big)\,\beta(v)} \qquad (6.28)$$

where $\psi(q)$ is initialized for each ending arc q_E by $\varphi(q_E) = 0$.

The recursive computation of $\varphi(q)$ in (6.27) is illustrated in Figure 6.4. Given the partial observation sequence $x_{r,1}, \ldots, x_{r,eq}$, $[\varphi(p) + C(q)]$ is the mean accuracy count of the sublattice that represents all partial paths that pass p and end with q. These paths are marked by the dotted line

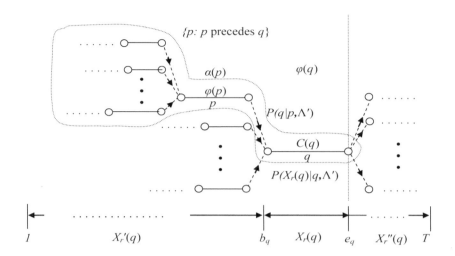

FIGURE 6.4: Illustrations of the sublattice containing arc q and of the recursive $\varphi(q)$ computation based on the sublattice. Each solid line represents an arc in the sublattice, and each dashed line represents the transition between two arcs. The dotted line encircles all partial paths that pass p and end with q.

in Figure 6.4. $\varphi(q)$ is a weighted sum and the weighted associated with each path passing arc p is $\alpha(p)P(q|p, \Lambda')p(X_r(q)|q, \Lambda')$, where each of the three factors is associated with each corresponding portion that makes up the path. The three factors are placed in the corresponding portions on the path in Figure 6.4. The weighted average of $[\varphi(p) + C(q)]$ over all arcs p (directly preceding q) using the three-factor weight above gives the recursive form of $\varphi(q)$ shown in the first line of (6.27). The second line of (6.27) removes some redundant computation and has been implemented in practice.

The recursive computation of $\psi(q)$ in (6.28) can be similarly interpreted as the weighted average of the accuracy count $C(v) + \psi(v)$ for all arcs v directly following q.

Now given that both $\varphi(q)$ and $\psi(q)$ are computed, and assuming that arc q depends only on the arcs directly preceding it and succeeding it, we can use (6.25) and (6.26) to directly prove that

$$\bar{C}_r(q) = \varphi(q) + \psi(q) \tag{6.29}$$

as one of the two quantities required to compute $\Delta\gamma(i, r, t)$ in (6.23). The interpretation of (6.29) is offered by using Figure 6.3. By definition, $\bar{C}_r(q)$ is the average accuracy count for utterance r over the sublattice shown in Figure 6.3 that contains arc q. This count can be decomposed into two parts. The first part is the "forward" average accuracy count of the left part of the sublattice in Figure 6.3 for the utterance from $t = 1$ to e_q, which is $\varphi(q)$. The second part is the "backward" average accuracy count of the right part of the sublattice for the utterance from $t = e_q + 1$ to T, which is $\psi(q)$.

The second quantity, $\bar{C}_r$, required to compute $\Delta\gamma(i, r, t)$ in (6.23) can be proved to be

$$\bar{C}_r = \frac{\displaystyle\sum_{q:q\in\{\text{ending arcs}\}} \varphi(q)\alpha(q)}{\displaystyle\sum_{q:q\in\{\text{ending arcs}\}} \alpha(q)} \tag{6.30}$$

The interpretation of (6.30) is as follows. Let arc q be an ending arc in the lattice. And recall that $\varphi(q)$ is the average accuracy count of utterance r given the sublattice that represents all s_r containing (sublattice-ending) arc q, and $\varphi(q)$ is the weight of this sublattice. Therefore, $\bar{C}_r$, which is defined in (6.22) as the average accuracy count for the entire lattice, becomes a weighted sum of the average accuracy counts of all sublattices as shown in (6.30).

This completes the description of the computation of $\Delta\gamma(i, r, t)$ in (6.23).

6.2.3 Computing $\Delta\gamma(i, r, t)$ for MCE Involving Lattices

Finally, we discuss using lattice approximation (6.13) to compute $\Delta\gamma(i, r, t)$ of (6.11) for MCE. As we mentioned earlier, whereas (6.13) is unified between MPE and MCE, the specific form of $C(s_r) = \delta(s_r, S_r)$ in MCE permits special simplification of $\Delta\gamma(i, r, t)$ of (6.11) for MCE. The simplification steps, followed by the use of (6.13), lead to

$$\Delta\gamma(i,r,t) = \sum_{s_r} p\left(s_r \middle| X_r, \Lambda'\right) \left[C(s_r) - p\left(S_r \middle| X_r, \Lambda'\right)\right] \gamma_{i,r,s_r}(t)$$

$$= p\left(S_r \middle| X_r, \Lambda'\right) \gamma_{i,r,S_r}(t) - p\left(S_r \middle| X_r, \Lambda'\right) \left[\sum_{s_r} p\left(s_r \middle| X_r, \Lambda'\right) \gamma_{i,r,s_r}(t)\right]$$

$$= p\left(S_r \middle| X_r, \Lambda'\right) \gamma_{i,r,S_r}(t) - p\left(S_r \middle| X_r, \Lambda'\right) \left[\sum_{q:t\in[b_q, e_q]} \gamma_{i,r,q}(t) \cdot \sum_{s_r:q\in s_r} p\left(s_r \middle| X_r, \Lambda'\right)\right]$$

$$= p\left(S_r \middle| X_r, \Lambda'\right) \gamma_{i,r,S_r}(t) - p\left(S_r \middle| X_r, \Lambda'\right) \left[\sum_{q:t\in[b_q, e_q]} \gamma_{i,r,q}(t) \cdot p\left(q \middle| X_r, \Lambda'\right)\right]$$

$$= p\left(S_r \middle| X_r, \Lambda'\right) \left[\underbrace{\gamma_{i,r,S_r}(t)}_{\gamma_{i,r}^{\text{num}}(t)} - \underbrace{\sum_{q:t\in[b_q, e_q]} \gamma_{i,r,q}(t) \cdot p\left(q \middle| X_r, \Lambda'\right)}_{\gamma_{i,r}^{\text{den}}(t)}\right] \tag{6.31}$$

The last line shows striking similarity between lattice-based MCE and MMI. In (6.31), $p(q \mid X_r, \Lambda') = \dfrac{p(q, X_r \mid \Lambda')}{p(X_r \mid \Lambda')}$ is computed by (6.16) and (6.17) for the numerator and denominator, respectively. Also in (6.31), we have $p(S_r \mid X_r, \Lambda') = \dfrac{p(X_r \mid S_r, \Lambda')p(S_r \mid \Lambda')}{p(X_r \mid \Lambda')}$, where correct string S_r is known. Hence, $\gamma_{i,r,S_r}(t)$ and $p(X_r \mid S_r, \Lambda')$ in (6.31) can be efficiently computed by the standard forward–backward algorithm for the HMM [43]. Finally, for the computation of $p(S_r \mid \Lambda')$ and $p(X_r \mid \Lambda')$, we use the language model and $\sum_{q:q \in \{\text{ending arcs}\}} \alpha(q)$, respectively.

Note that the computation for the lattice-based MCE we provided in (6.31) does not require removing the correct word string S_r from the lattice.

6.3   ARBITRARY EXPONENT SCALING IN MCE IMPLEMENTATION

In this section, we discuss one of the two empirical issues in MCE implementation that were raised in Chapter 3. In (3.15), if we use the exponent scaling factor $\eta \neq 1$, we can obtain the following result corresponding to (3.17):

$$l_r(d_r(X_r, \Lambda)) = \frac{\sum\limits_{s_r, s_r \neq S_r} p^\eta(X_r, s_r \mid \Lambda)}{\sum\limits_{s_r} p^\eta(X_r, s_r \mid \Lambda)}$$

The corresponding result to (3.19) then becomes

$$O_{\text{MCE}}(\Lambda) = \sum_{r=1}^R \frac{p^\eta(X_r, S_r \mid \Lambda)}{\sum\limits_{s_r} p^\eta(X_r, s_r \mid \Lambda)}$$

which can be reformulated into a rational function using the same steps as in Section 3.4.2:

$$O_{\text{MCE}}(\Lambda) = \frac{\sum\limits_{s_1, \dots, s_R} p^\eta(X_1, \dots, X_R, s_1, \dots, s_R \mid \Lambda) \, C_{\text{MCE}}(s_1, \dots, s_R)}{\sum\limits_{s_1, \dots, s_R} p^\eta(X_1, \dots, X_R, s_1, \dots, s_R \mid \Lambda)} \tag{6.32}$$

The remaining derivations in Chapters 4 and 5 will no longer follow strictly for the more general and practical case of (6.32). In the MCE implementation that we have done, however, we modify (6.11) for computing $\Delta\gamma(i, r, t)$ in the following manner in order to include the effects of the exponent scaling factor:

$$\Delta\gamma(i, r, t) = \sum_{s_r} \tilde{p}(s_r \mid X_r, \Lambda') \left(C(s_r) - \sum_{s_r} \tilde{p}(X_r \mid s_r, \Lambda') C(s_r) \right) \gamma_{i, r, s_r}(t) \tag{6.33}$$

where $\tilde{p}(s_r|X_r, \Lambda')$ is the generalized posterior probability of s_r, which can be computed as

$$\tilde{p}\left(s_r\middle|X_r, \Lambda'\right) = \frac{p^\eta\left(X_r, s_r\middle|\Lambda'\right)}{\displaystyle\sum_{s_r} p^\eta\left(X_r, s_r\middle|\Lambda'\right)} \tag{6.34}$$

After this modification, all derivations in Chapters 4 and 5 are unchanged.

6.4 ARBITRARY SLOPE IN DEFINING MCE COST FUNCTION

The second empirical MCE implementation issue raised in Chapter 3 concerns the use of $\alpha \neq 1$ in (3.16). For 1-best MCE, α acts as η; that is, we can equivalently set $\eta = \alpha$, and $\alpha = 1$. Then, we can compute $\Delta\gamma(i, r, t)$ according to (6.33). For N-best MCE ($N > 1$), given the discriminant function defined in (3.15) and sigmoid function defined in (3.16), we have the following result corresponding to (3.17):

$$l_r\left(d_r(X_r, \Lambda)\right) = \frac{\left(\displaystyle\sum_{s_r,\, s_r \neq S_r} p^\eta\left(X_r, s_r\middle|\Lambda\right)\right)^\alpha}{p^{\eta\cdot\alpha}\left(X_r, S_r\middle|\Lambda\right) + \left(\displaystyle\sum_{s_r,\, s_r \neq S_r} p^\eta\left(X_r, s_r\middle|\Lambda\right)\right)^\alpha} \tag{6.35}$$

Now, α is applied outside of the summation of scaled joint probabilities over all competing strings, making rigorous computation intractable. In our practical MCE implementation, we instead use $\sum_{s_r,\, s_r \neq S_r} p^{\alpha\cdot\eta}(X_r, s_r|\Lambda)$ to approximate $(\sum_{s_r,\, s_r \neq S_r} p^\eta(X_r, s_r|\Lambda))^\alpha$. This approximation (which is exact when η approaches infinity) makes it equivalent to setting the new "η" as $\alpha \cdot \eta$, and setting the new $\alpha = 1$. Then, again, we can compute $\Delta\gamma(i, r, t)$ according to (6.33). It should be noted that, with this approximation, the computation for the lattice-based MCE we provided in (6.31) does not require removing the correct word string S_r from the lattice, as shown in (6.31) (Section 6.2.3). This contrasts the solution in [31, 46] where the removal was necessary without using the approximation, making it more difficult to implement in practice.

The two empirical solutions cited above have been successfully implemented in our speech recognition system, yielding strong practical results (published in [20, 58]) that validate the solutions.

·   ·   ·   ·   ·

CHAPTER 7

Selected Experimental Results

In this chapter, we present experimental results on several automatic speech recognition (ASR) tasks. We evaluate the growth transformation (GT)-based minimum classification error (MCE) training method on both small-vocabulary, well-controlled benchmark tests such as TIDIGITS, and on large-vocabulary, real-world speech recognition tasks such as commercial telephony large-vocabulary ASR (LV-ASR) applications. We show that the GT-based discriminative training gives superior performance over the conventional maximum likelihood (ML)-based training method.

7.1   EXPERIMENTAL RESULTS ON SMALL ASR TASKS TIDIGITS

TIDIGITS is a speaker-independent connected-digit task. Each utterance in this corpus has an unknown length with a maximum of seven digits. The training set includes 8623 utterances and testing set includes 8700 utterances. In our experiment, a word-based hidden Markov model (HMM) is built for each of the 10 digits from ZERO to NINE, plus word OH. The number of states of each HMM ranges from 9 to 15, depending on the average duration of each word, and each state has an average of six Gaussian mixture components. The speech feature vector is computed from audio signal analysis, which gives 12 Mel frequency cepstral coefficients (MFCCs) and the audio energy, plus their first-order and second-order temporal differences.

GT-MCE training is performed with initialization from the ML-trained models. In this experiment, 1-best MCE training is used. The constant factor E as discussed in Section 5.3 is set to 1.0, and the sloping factor α is set to 0.01.

The experimental results are shown in Table 7.1. The baseline HMMs are trained with the ML criterion. The ML model gives a word error rate (WER) of 0.30% on testing data. With GT/extended Baum–Welch (EBW)-based MCE training, after three iterations, the algorithm convergence is reached and the WER is reduced to 0.23%. As a comparison, a conventional generalized probabilistic descent (GPD)-based MCE is also implemented for this task. As shown in Table 7.1, the best GPD MCE result is with a WER of 0.24%, which is obtained after 12 iterations over the full training data set (i.e., 12 epochs). The results of this small-task experiment show that the new GT-based MCE learning method is slightly better than the conventional GPD-based MCE,

TABLE 7.1: Comparative recognition-accuracy performance (measured by WER — the lower the better) of the new and traditional MCE training methods, as well as the ML method

	ML	GPD MCE	GT/EBW MCE
WER	0.30%	0.24%	0.23%
WER reduction	–	20.0%	23.3%

and it gives significantly improved efficiency in the training by providing much faster algorithm convergence.

Figure 7.1 gives a detailed analysis of the GT-based MCE learning. Figure 7.1a shows that the value of the objective function of MCE training decreases monotonically for each new iteration. Note that the loss function of the training set is the one defined by (3.18). Figure 7.1b shows that the WER on the test set is reduced significantly also after a few GT iterations. It is noteworthy that given different settings of the global constant E as discussed in Chapter 5.3, the model updating speed can be controlled. For example, a smaller E leads to faster training speed. However, the models trained this way may suffer from overtraining and hence cause unstable performance on the test set.

7.2   TELEPHONY LV-ASR APPLICATIONS

We further evaluated the GT-based discriminative training method on large vocabulary telephony speech recognition tasks. The Microsoft large-scale telephony speech databases are used to build a large-vocabulary telephony ASR system. The entire training set, which is collected through various channels including close-talk telephones, far-field microphones, and cell phones, consists of 26 separate corpuses, 2.7 million utterances, and a total of 2000 hours of speech data. To improve the robustness of the acoustic model, speech data are recorded under various conditions with different environmental noises and include both native English speakers and speakers with various foreign accents. The text prompts include common telephony-application style utterances and some dictation-style utterances from the *Wall Street Journal* database.

The test sets consist of several typical context-free grammar (CFG)-based commercial telephony ASR tasks. To evaluate the generalization ability of our approach, the test data are collected in a very different setup than the training set. The default global vocabulary size of the ASR system is 120K. However, different vocabularies are used in different test sets. Table 7.2 summarizes the test sets used in our experiments.

In this experiment, all data are sampled at a rate of 8K Hz. Phonetic decision trees are used for state tying and there are about 6000 tied states with an average of 16 Gaussian mixture components per state. The 52-dimensional raw acoustic feature vectors are composed of the normalized energy,

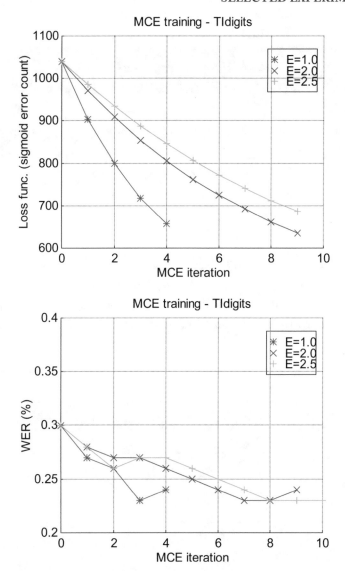

FIGURE 7.1: (a) GT-based MCE training. Number of iterations versus loss function of the training set given different settings of the global constant E that controls the model updating speed. (b) GT-based MCE training. Number of iterations versus WER on the test set given different settings of the global constant E that controls the model updating speed.

and 12 MFCCs and their first-, second-, and third-order time derivatives. The 52-dimensional raw features are further projected to form 36-dimensional feature vectors via heteroscedastic linear discriminant analysis transformation.

TABLE 7.2: Description of the test sets			
NAME	VOCABULARY SIZE	WORD COUNT	DESCRIPTION
MSCT	70K	4356	General call center application.
STK	40K	12,851	Finance applications (stock transaction, etc.)
QSR	55K	5718	Name dialing application (note: pronunciations of most names are generated by letter-to-sound rules).

MSCT, Microsoft Connect (an MS internal data set); STK, Stock (an MS internal data set about stack query); QSR, QuickSilver (an MS internal data set).

As with the TIDIGITS database, GT-MCE training is performed with initialization from the ML-trained models. The sloping facto α is set to 1/30 and the E factor is set to 1.0. In MCE training of HMMs, the training data are first decoded by a simple unigram weighted CFG containing all the words in the transcripts of training data and the competitors are then updated after each

TABLE 7.3: Experimental results on three telephony ASR test sets			
	TEST SET	ML	MCE
MSCT	WER	12.413%	10.514%
	Abs. WERR Over ML	N/A	1.899%
	Rel. WERR Over ML	N/A	15.30%
STK	WER	7.993%	7.330%
	Abs. WERR Over ML	N/A	0.663%
	Rel. WERR Over ML	N/A	8.30%
QSR	WER	9.349%	8.464%
	Abs. WERR Over ML	N/A	0.885%
	Rel. WERR Over ML	N/A	9.47%
Average	WER	9.918%	8.769%
	Abs. WERR Over ML	N/A	1.149%
	Rel. WERR Over ML	N/A	11.58%

N/A, Not Applied; Rel., Relative; Abs., Absolute

set of three iterations. All HMM parameters (except transition probabilities and Gaussian mixture weights) are updated. To prevent variance underflow, a dimension-dependent variance floor is set as 1/20 of the average variance over all Gaussian components in that dimension. The variance values that are lower than the variance floor are set to the floor value. In the experiments, only six iterations (i.e., six epochs) are performed in MCE training.

Table 7.3 presents the WER on the three test sets. Compared with the ML baseline, the MCE training can reduce the WER by 11.58%. These results demonstrate that the MCE training approach has strong generalization ability. It can be effectively applied not only to small-scale tasks but also to large-scale ASR tasks.

CHAPTER 8

Epilogue

8.1 SUMMARY OF BOOK CONTENTS

This book starts by providing an introduction to discriminative learning, speech recognition, and the roles of discriminative learning in speech recognition. Then it presents some background material on basic probability distributions and on optimization techniques; both serve as mathematical pre-requisites for the remaining book content dealing with detailed techniques for discriminative learning in speech recognition. The basic probability distributions covered in the background material include multinomial distribution and multivariate Gaussian distribution, both belonging to the more general exponential-distribution family, as well as Gaussian mixture distribution, which is outside of the exponential-distribution family. The optimization concepts and techniques covered in the background material include definitions of global and local optimums, their necessary condition, Lagrange multiplier method, gradient-based method, and, finally, growth transformation (GT) method.

The book then proceeds by a tutorial on statistical speech recognition, where the hidden Markov model (HMM) is formally introduced as a popular acoustic model for speech feature sequences, and the language model is also introduced as the prior probability of word sequences. Introduction of the HMM sets up the context in which discriminative learning, as the main subject of this book, is subsequently discussed in a great detail.

Given the concepts of acoustic modeling and language modeling, we then provide a unified account for three common objective functions for discriminative training of HMMs currently in use in speech recognition practice, including maximum mutual information (MMI), minimum classification error (MCE), and minimum phone error/minimum word error (MPE/MWE). Comparisons are made between our unified form of the objective functions with another unified form of discriminative objective function in the literature and insights are offered in the comparisons.

The subsequent materials covered in this book focus on ways of carrying out discriminative parameter learning using the unified form of objective functions via the GT technique. The coverage starts with a relatively simple case where the IID assumption for data observation (or model stationarity) is made and where exponential-family distributions are assumed for the data. Next, the more complex, nonstationary model for sequentially correlated data observations is discussed.

HMMs are used in this case and its parameter estimation problem via the GT technique is presented in detail.

Some practical implementation issues of the GT technique for HMM parameter learning are then discussed, filling in some details that were not dealt with in the preceding portions so as not to unnecessarily divert the main topics. Then, finally, we present some selected experimental results in speech recognition, demonstrating the effectiveness of the techniques presented in this book in practice.

8.2   SUMMARY OF CONTRIBUTIONS

The main technical contribution of this book is to provide three aspects of unification of the common discriminative learning techniques for sequential pattern recognition, in particular, those for speech recognition.

- First, the unification is in the objective function for optimization, which has been derived rigorously to be a rational functional form of (3.2). Although the rational functional form for MMI has been known in the past, we provide the first proof that the same form applies to MCE and MPE/MWE, differing from MMI only in the constant weighting factors.
- Second, the unification is in the optimization technique. The unified, rational functional form of the objective function for MMI, MCE, and MPE/MWE enables the use of the special technique of GT/extended Baum–Welch for optimization. In the past, MCE had always been implemented by gradient descent, due to the lack of any rational functional form in its objective function. The essence of the GT technique is to optimize the specially constructed auxiliary function, for both the discrete valued HMM and for the continuous valued Gaussian HMM.
- Third, the unification is in the final parameter reestimation formulas. The formula for the HMM transition probability is shown to be (5.26). The formula for the discrete HMM's emitting probability is shown to be (5.21). The formulas for the Gaussian HMM is shown to be (5.35) and (5.36) for its mean vectors and covariance matrices, respectively.

The unifying review of discriminative learning for HMMs provided in this book is motivated by the striking success of such various techniques in recent speech recognition research and system development. Yet, there has been a conspicuous lack of common understanding of the relationships among these techniques including MMI, MCE, and MPE/MWE, despite the relatively long history of MMI (since 1987 [6]), MCE (since 1992 [24]), and MPE/MWE (since 2002 [38]).

Because of the complexity of these techniques and the lack of a unifying theoretical theme underlying them, only a very small number of speech recognition laboratories worldwide have been able to successfully implement these techniques and to achieve similarly strong performance gains for large vocabulary speech recognition. The main goal of this book is to provide just such a unifying theoretical framework in hopes of promoting more widespread use of discriminative learning not only in speech recognition, but possibly in other types of sequential pattern recognition and signal processing problems as well. It is also hoped that given a solid theoretical foundation presented in this book, other more advanced pattern recognition concepts (e.g., discriminative margins [48]) can be more elegantly integrated with current discriminative learning techniques. The new goal then is not only to reduce empirical errors but also to enhance generalization capabilities.

In this book, we show in a step-by-step fashion that our approach leads to consistent parameter estimation results and it is scalable for large-scale pattern recognition tasks. We also analyzed the algorithmic properties of the MCE- and MPE/MWE-based learning methods under the GT parameter estimation framework for sequential pattern recognition using HMMs.

8.3   REMAINING THEORETICAL ISSUE AND FUTURE DIRECTION

The material covered in this book is probably the most comprehensive one on the topic of discriminative learning designed for sequential pattern recognition such as speech recognition. One important theoretical issue for the CDHMM concerns the convergence properties of the GT method. In Section 5.3, we discussed this issue in depth, where we outlined a proof (based on Axelrod et al.'s work [3]) that the GT update formulas for the CDHMM are valid given a sufficiently large (but bounded) constant D_i. However, in that analysis, no explicit construction of D_i was given. Therefore, it constitutes only an existence proof. One interesting remaining issue is whether one can provide a constructive proof. In this final section, we outline one constructive proof for advanced readers, based on Jebera's work [22, 23], where reverse Jensen's inequality was used for optimization.

In principle, Jebara's method is applicable to maximizing any rational function, whose numerator and denominator can be mixtures of exponential models. Therefore, it is applicable to optimizing our unified discriminative criterion (3.2) for all MMI, MCE, and MPE/MWE. In the brief review below, we introduce the principle of reverse Jensen's inequality and its application to discriminative objective function optimization.

For a rational function in the form of (1.26) and (1.27), we desire to maximize the following equivalent function:

$$\log O(\Lambda) = \log G(\Lambda) - \log H(\Lambda) = \log \sum_s p(X, s | \Lambda) \, C(s) - \log \sum_s p(X, s | \Lambda) \qquad (8.1)$$

The first term in (8.1) is a log-sum function similar to log likelihood. Based on the well-known Jensen's inequality and after several steps of simplifications, we have

$$\log \sum_{s} p(X,s|\Lambda)\,C(s) \geq Q_G(\Lambda;\Lambda') + J$$

$$= \underbrace{\sum_{s} \left(\frac{p(X,s|\Lambda)\,C(s)}{\sum_{s} p(X,s|\Lambda')\,C(s)} \right) \log p(X,s|\Lambda)\,C(s) + J}_{Q_G(\Lambda;\Lambda')} \qquad (8.2)$$

where J is a constant irrelevant to Λ (although relevant to Λ'), that is, $J = \log\sum_{s} p(X, s|\Lambda')C(s) - Q_G(\Lambda'; \Lambda')$. This is similar to the E-step in the EM algorithm.

The left-hand side of (8.2) is a lower bound of $\log G(\Lambda)$, and makes tangential contact with $\log G(\Lambda)$ at Λ'. Therefore maximizing the auxiliary function $Q_G(\Lambda; \Lambda')$ guarantees increase of $\log G(\Lambda)$ iteratively.

However, to maximize $\log O(\Lambda)$, we need a lower bound for $\log O(\Lambda)$, which in turn requires an upper bound of $\log H(\Lambda)$. In [22], it was shown (nontrivially) that using reverse Jensen's inequality an auxiliary function $Q_H(\Lambda; \Lambda')$ can be constructed so that

$$\log \sum_{s} p(X,s|\Lambda) \leq Q_H(\Lambda;\Lambda') + \tilde{J} = \underbrace{\sum_{s}(-w_s)\log p(Y_s,s|\Lambda) + \tilde{J}}_{Q_H(\Lambda;\Lambda')} \qquad (8.3)$$

where $\tilde{J}$ is a Λ-irrelevant constant that makes $Q_H(\Lambda; \Lambda') + \tilde{J}$ tangential contact with $\log H(\Lambda)$ at Λ', and w_s and Y_s are positive weights and modified observations, respectively. Reverse Jensen's inequality was derived by exploiting the convexity of the cumulant generating function of exponential family in [22] and will not be elaborated further here.

Given (8.2) and (8.3), one can construct the auxiliary function for $\log O(\Lambda)$ as:

$$Q(\Lambda;\Lambda') = Q_G(\Lambda;\Lambda') - Q_H(\Lambda;\Lambda') \qquad (8.4)$$

where $Q(\Lambda; \Lambda')$ is a lower bound of $\log O(\Lambda)$ and makes tangential contact with $\log O(\Lambda)$ at Λ'. Therefore, optimizing $\log O(\Lambda)$ can be achieved by iteratively optimizing $Q(\Lambda; \Lambda')$, which takes the same step as the M-step in the conventional EM algorithm for an HMM (i.e., with a closed-form solution in the M-step).

Note for our unifying discriminative objective function (3.2), the summand of $G(\Lambda)$ may take a negative value for MPE; that is, for some paths s that have many insertion errors, the corresponding $C(s)$ may be negative. In this case, we can add extra dummy training tokens to the training set, while these dummy tokens can only be recognized as correct references. Appending these dummy tokens to s can effectively increase its raw accuracy count to be positive. Moreover, because the dummy token will not compete with any other tokens in the training set, it will not affect the training performance.

Applications of reverse Jensen's inequality for discriminative training have been an interesting research area recently. In [1], the method for MMI optimization was explored and was compared with the conventional GT method using empirical setting of the constant D_i. After approximation and simplification, the author showed similar forms of model estimation formulas to those derived from the empirical GT method, but with a larger D_i, and slower convergence. Further investigation of the method based on reverse Jensen's inequality for discriminative training is warranted, and this constitutes one fruitful theoretical research direction for full maturity of learning algorithms for discriminative training in the statistical recognizer design.

Major Symbols Used in the Book and Their Descriptions

Symbols	DESCRIPTIONS
R	number of training samples (tokens or strings)
$r = 1, \ldots, R$	index of individual training samples
X	aggregate of all training samples
χ	a random variable in the space of X
S	aggregate of reference labels in the training data
s	aggregate of hypothesis labels (reference or otherwise)
X_r	rth training sample. It may represent a sequence of observation vectors with a variable length: $X_r = x_{r,1}, x_{r,2}, \ldots, x_{r,T}$
χ_r	a random variable in the space of X_r
S_r	reference label of the rth training sample. It may represent a sequence of words with variable length: $S_r = w_{r,1}, w_{r,2}, \ldots, w_{r,M}$
s_r	hypothesis label of the rth training sample (reference or otherwise)
$x_{r,t}$	tth feature vector of the rth training observation sequence.
$\chi_{r,t}$	A random variable in the space of $x_{r,t}$.
$w_{r,i}$	ith word of the reference label of the rth training sample
q	hidden Markov model (HMM) state sequence
θ	natural parameters of the exponential family distributions
Λ	Aggregate of model parameter sets
Λ'	model parameter sets obtained from the immediately previous iteration in an iterative learning algorithm
$a_{i,j}$	HMM transition probability from state i to state j
$b_i(x)$	HMM emitting probability for observation x at state i
μ	mean vector of Gaussian distribution
Σ	covariance matrix of Gaussian distribution

Mathematical Notation

1. Superscript T denotes the transpose of a matrix or vector; for example, $\mathbf{x}^\mathrm{T}$ will be a row vector.

2. $(w_1, \ldots, w_M)$ denotes a row vector with M elements, and the corresponding column vector is denoted as $\mathbf{w} = (w_1, \ldots, w_M)^\mathrm{T}$.

3. $[a, b]$ denotes the closed interval from a to b (i.e., the interval including the values a and b themselves). (a, b) denotes the corresponding open interval (i.e., the interval excluding a and b). $[a, b)$ denotes an interval that includes a but excludes b.

4. The expectation of a function $f(x)$ with respect to a random variable x is denoted by $\mathbb{E}_{p(x|\lambda)}[f(x)]$ or $\mathbb{E}_x[f(x)]$, assuming that the distribution of x is described by $p(x|\lambda)$, where λ is the parameter set in this distribution. In situations where there is no ambiguity on which a variable is being averaged over, this will be simplified by removing the suffix.

Bibliography

[1] M. Afify, "Extended Baum–Welch re-estimation of Gaussian mixture models based on reverse Jensen's inequality," *Proc. INTERSPEECH*, 2005, 1113–1116.

[2] S. Amari, "A theory of adaptive pattern classifiers," *IEEE Trans. Electron. Comput.*, 16, 1967, 299–307.

[3] S. Axelrod, V. Goel, R. Gopinath, P. Olsen, and K. Visweswariah, "Discriminative estimation of subspace constrained Gaussian mixture models for speech recognition," *IEEE Trans. Audio, Speech Lang. Proc.*, 15, 2007, 172–189.

[4] L. Baum and G. Sell, "Growth transformations for functions on manifolds," *Pac. J. Math.*, 27(2), 1968, 211–227.

[5] E. Birney, "Hidden Markov models in biological sequence analysis," *IBM J. Res. Dev.*, 45, 2001, 449–454.

[6] P. Brown, The acoustic modeling problem in automatic speech recognition, Ph.D. thesis, Carnegie Mellon University, 1987.

[7] W. Chou and L. Li, "A minimum classification error framework for generalized linear classifier in machine learning and its application to text categorization/retrieval," *Proc. Int. Conf. Mach. Learning Appl.*, Dec. 2004, 382–390.

[8] W. Chou, "Minimum classification error approach in pattern recognition," in W. Chou and B.-H. Juang (eds.), *Pattern Recognition in Speech and Language Processing*, CRC Press, Boca Raton, 2003, pp. 1–49.

[9] M. Collins, "Discriminative training methods for hidden Markov models: theory and experiments with perceptron algorithms," *Proc. Empirical Methods of Natural Language Processing (EMNLP)*, 2005, 1–8.

[10] L. Deng and D. O'Shaughnessy, *Speech Processing — A Dynamic and Optimization-Oriented Approach*, Marcel Dekker Publishers, New York, NY, 2003.

[11] L. Deng, D. Yu, and A. Acero, "A generative modeling framework for structured hidden speech dynamics," Neural Information Processing System (NIPS) Workshop, Whistler, BC, Canada, Dec. 2005.

[12] L. Deng, J. Wu, J. Droppo, and A. Acero, "Analysis and comparison of two feature extraction/compensation algorithms," *IEEE Signal Process. Lett.*, 12(6), June, 2005, 477–480.

[13]   R. Durbin, S. Eddy, A. Krogh, and G. Mitchison, *Biological Sequence Analysis: Probabilistic Models of Proteins and Nucleic Acids*, Cambridge University Press, 1998, 1–298.

[14]   P. Gopalakrishnan, D. Kanevsky, A. Nadas, and D. Nahamoo, "An inequality for rational functions with applications to some statistical estimation problems," *IEEE Trans. Inf. Theory*, 37, January 1991, 107–113. doi:10.1109/18.61108

[15]   Y. Gao and J. Kuo, "Maximum entropy direct models for speech recognition," *IEEE Trans. Speech Audio Process.*, 2006.

[16]   V. Goel and W. Byrne, "Minimum Bayes risk methods in automatic speech recognition," in W. Chou and B.-H. Juang (eds.), *Pattern Recognition in Speech and Language Processing*, CRC Press, Boca Raton, 2003, 51–80.

[17]   A. Gunawardana and W. Byrne, "Discriminative speaker adaptation with conditional maximum likelihood linear regression," *Proc. EUROSPEECH*, 2001.

[18]   A. Gunawardana, M. Mahajan, A. Acero, and J. Platt, "Hidden conditional random fields for phone classification," *Proc. Interspeech*, Lisbon, 2005.

[19]   X. He and W. Chou, "Minimum classification error linear regression for acoustic model adaptation of continuous density HMMs," *Proc. ICASSP*, April 2003.

[20]   X. He, L. Deng, and W. Chou, "A novel learning method for hidden Markov models in speech and audio processing," *Proc. IEEE Workshop on Multimedia Signal Processing*, Victoria, BC, October 2006.

[21]   X. Huang, A. Acero and H. Hon, *Spoken Language Processing*, Prentice Hall, Upper Saddle River, New Jersey, 2001.

[22]   T. Jebara, Discriminative, Generative and imitative learning, PhD Thesis, Media Laboratory, MIT, December 2001.

[23]   T. Jebara and A. Pentland, "On reversing Jensen's inequality," *Neural Inf. Process. Syst.*, 13, December 2000, 231–237.

[24]   B.-H. Juang and S. Katagiri, "Discriminative learning for minimum error classification, *IEEE Trans. Signal Process.*, 40(12), 1992, 3043–3054. doi:10.1109/78.175747

[25]   B.-H. Juang, W. Chou, and C.-H. Lee, "Minimum classification error rate methods for speech recognition," *IEEE Trans. Speech Audio Process*, 5, May 1997, 257–265.

[26]   D. Kanevsky, "Extended Baum transformations for general functions," *Proc. ICASSP*, 2004. doi:10.1109/ICASSP.2004.1326112

[27]   D. Kanevsky, "A generalization of the Baum algorithm to functions on non-linear manifolds," *Proc. ICASSP*, 1995. doi:10.1109/ICASSP.1995.479631

[28]   J. Lafferty, A. McCallum, and F. Pereira, "Conditional random fields: Probabilistic models for segmenting and labeling sequence data," *Proc. Int. Conf. Mach. Learning*, 2001, 282–289.

[29] Y. Li, L. Shapiro, and J. Bilmes, "A generative/discriminative learning algorithm for image classification," *IEEE Int. Conf. Computer Vision*, Beijing, China, 2005, 1605–1612.

[30] A. McCallum, D. Freitag, and F. Pereira, "Maximum entropy Markov models for information extraction and segmentation." *Proc. Int. Conf. Mach. Learning*, 2000, 591–598.

[31] W. Macherey, L. Haferkamp, R. Schlüter, and H. Ney, "Investigations on error minimizing training criteria for discriminative training in automatic speech recognition," *Proc. Interspeech*, 2005, Lisbon, 2133–2136.

[32] E. McDermott, "Discriminative training for automatic speech recognition using the minimum classification error framework," *Neural Inf. Process. Syst.* (NIPS) Workshop, Whistler, BC, Canada, Dec. 2005.

[33] E. McDermott, T. Hazen, J. L. Roux, A. Nakamura, and S. Katagiri, "Discriminative training for large vocabulary speech recognition using minimum classification error," *IEEE Trans. Speech Audio Process.*, 15(1), 2007, 203–223.

[34] Y. Normandin, Hidden Markov models, maximum mutual information estimation, and the speech recognition problem, PhD. dissertation, McGill University, Montreal, 1991.

[35] Y. Normandin, "Maximum mutual information estimation of hidden Markov models," in C.-H. Lee, F.K. Soong, and K.K. Paliwal (eds.), *Automatic Speech and Speaker Recognition*, Kluwer Academic Publishers, Norwell, MA, 1996, 1–159.

[36] F. Och, "Minimum error rate training in statistical machine translation," *Proc. 41st Meeting Assoc. Comp. Linguistics*, 2003, 160–167. doi:10.3115/1075096.1075117

[37] F. Pereira, "Linear models for structure prediction," *Proc. Interspeech*, Lisbon, 2005, 717–720.

[38] D. Povey and P.C. Woodland, "Minimum phone error and I-smoothing for improved discriminative training," *Proc. ICASSP*, 2002. doi:10.1109/ICASSP.2002.1005687

[39] D. Povey, M.J.F. Gales, D.Y. Kim, and P.C. Woodland, "MMI-MAP and MPE-MAP for acoustic model adaptation," *Proc. Eurospeech*, 2003.

[40] D. Povey, Discriminative training for large vocabulary speech recognition, Ph.D. dissertation, Cambridge University, Cambridge, UK, 2004.

[41] D. Povey, B. Kingsbury, L. Mangu, G. Saon, H. Soltau, and G. Zweig, "fMPE: discriminatively trained features for speech recognition," *Proc. DARPA EARS RT-04 Workshop*, Nov. 7–10, 2004, Palisades, NY, Paper No. 35, 5 pp.

[42] C. Rathinavalu and L. Deng, "Speech trajectory discrimination using the minimum classification error learning," *IEEE Trans. Speech Audio Process.*, 6, 1998, 505–515.

[43] L. Rabiner and B.-H. Juang, *Fundamentals of Speech Recognition*, Prentice Hall, 1993, 1–507.

[44] J. Roux and E. McDermott, "Optimization for discriminative training," *Proc. INTERSPEECH*, Lisbon, 2005.

[45] R. Schlüter, Investigations on discriminative training criteria, Ph.D. dissertation, RWTH Aachenm University of Technology, Aachen, Germany, 2000.

[46] R. Schlüter, W. Macherey, B. Muller, and H. Ney, "Comparison of discriminative training criteria and optimization methods for speech recognition," *Speech Commun.*, 34, 2001, 287–310. doi:10.1016/S0167-6393(00)00035-2

[47] S. Theodoridis and K. Koutroumbas, *Pattern Recognition*, Elsevier Science, Academic Press, 2003, 1–625.

[48] V. Vapnik, *Statistical Learning Theory*, Wiley-Interscience, 1998, 1–768.

[49] V. Valtchev, Discriminative methods for HMM-based speech recognition, Ph.D. thesis, Cambridge University, England, 1995.

[50] V. Valtchev, J.J. Odell, P.C. Woodland, and S.J. Young, "MMIE training of large vocabulary recognition systems," *Speech Commun.*, 22(4), 1997, 303–314. doi:10.1016/S0167-6393(97)00029-0

[51] V. Valtchev, J.J., Odell, P.C., Woodland, and S.J. Young, "Lattice-based discriminative training for large vocabulary speech recognition," *Proc. ICASSP*, 1996.

[52] P. Woodland and D. Povey, "Large scale discriminative training for speech recognition," *Proc. ITRW ASR*, 2000. doi:10.1006/csla.2001.0182

[53] Y. Xiong, Q. Huo, C. Chan, "A discrete contextual stochastic model for the offline recognition of handwritten Chinese characters," *IEEE Trans. PAMI*, 23, 2001, 774–782.

[54] R. Yan, J. Zhang, J. Yang, and A. Hauptmann, "A discriminative learning framework with pairwise constraints for video object classification," *IEEE Trans. PAMI*, 28(4), 2006, 578–593.

[55] J. Yang, Y. Xu, and C.S. Chen, "Hidden Markov model approach to skill learning and its application in telerobotics," *IEEE Trans. Robotics Autom.*, 10(5), 1994, 621–631.

[56] C. Yen, S. Kuo, and C.-H. Lee, "Minimum error rate training for HMM-based text recognition," *IEEE Trans. Image Process.*, 8, 1999, 1120–1124.

[57] D. Yu, L. Deng, and A. Acero, "A* lattice search algorithm for a long-contextual-span hidden trajectory model and phonetic recognition," *Proc. Interspeech*, Lisbon, 2005, 553–556.

[58] D. Yu, L. Deng, X. He, A. Acero, "Large-margin minimum classification error training for large-scale speech recognition tasks," *Proceedings of ICASSP*, Honolulu, Hawaii, April 2007.

Author Biography

Xiaodong He received his bachelor's degree from Tsinghua University, Beijing, China, in 1996, and earned his master's degree from the Chinese Academy of Sciences in 1999, and his doctoral degree from the University of Missouri–Columbia in 2003. He joined the Speech and Natural Language group of Microsoft in 2003, and the Natural Language Processing group of Microsoft Research, Redmond, WA, in 2006, where he currently serves as researcher. His research areas include statistical machine learning, automatic speech recognition, natural language processing, machine translation, signal processing, nonnative speech processing, and human–computer interaction. In these areas, he has authored/coauthored more than 30 refereed papers in leading international conferences and journals. He has filed more than 10 U.S. or international patents in the areas of speech recognition, language processing, and machine translation. He served as a reviewer for major conferences and journals in the areas of speech recognition, natural language processing, signal processing, and pattern recognition. He also served on program committees of various conferences in these areas. He is a member of ACL, IEEE, ISCA, and Sigma Xi.

Li Deng received his bachelor's degree from the University of Science and Technology of China and his Ph.D. degree from the University of Wisconsin–Madison. In 1989, he joined the Department of Electrical and Computer Engineering, University of Waterloo, Ontario, Canada, as assistant professor; he became tenured full professor in 1996. From 1992 to 1993, he conducted sabbatical research at the Laboratory for Computer Science, Massachusetts Institute of Technology, Cambridge, MA, and from 1997 to 1998, at the ATR Interpreting Telecommunications Research Laboratories, Kyoto, Japan. During 1989–1999, he taught a wide range of electrical and computer engineering courses, both at undergraduate and graduate levels. In 1999, he joined Microsoft Research, Redmond, WA, as senior researcher; he currently serves as principal researcher for the same institution. He has also been affiliate professor in the Department of Electrical Engineering at University of Washington since 2000 after moving to Seattle. His past and current research areas include automatic speech and speaker recognition, statistical methods and machine learning, neural information processing, machine intelligence, audio and acoustic signal processing, statistical signal processing and digital communication, human speech production and perception, acoustic phonetics, auditory speech processing, noise robust speech processing, speech synthesis and enhancement, spoken language understanding systems, multimedia signal processing, and multimodal human–computer interaction. In these areas, he has published more than 300 refereed papers in leading international conferences and journals, and 14 book chapters, and has given keynotes, tutorials, and lectures worldwide. He has been granted more than 20 U.S. or international patents in acoustics, speech/language technology, and signal processing. He has likewise authored two recent books on speech processing.

He served on Education Committee (as a founding member) and Speech Processing Technical Committee of the IEEE Signal Processing Society 1996–2000, and was associate editor of *IEEE Trans. Speech and Audio Processing* for 2002–2005. He currently serves on the Society's Multimedia Signal Processing Technical Committee and on the Society's Board of Governors. He is editor-in-chief of the *IEEE Signal Processing Magazine* and associate editor of *IEEE Signal Processing Letters*. He was a technical chair of the 2004 IEEE International Conference on Acoustics, Speech, and Signal Processing (ICASSP04), and the general chair of the 2006 IEEE Workshop on Multimedia Signal Processing. He will be general chair of the ICASSP-2013 in Vancouver, Canada. He is a Fellow of the Acoustical Society of America and a Fellow of the IEEE.

• • • • •

Printed in the United States
by Baker & Taylor Publisher Services